Creating the Molecules of Life

AAS | IOP Astronomy

About the program:

AAS-IOP Astronomy ebooks is the official book program of the American Astronomical Society (AAS), and aims to share in depth the most fascinating areas of astronomy, astrophysics, solar physics and planetary science. The program includes publications in the following topics:

- GALAXIES AND COSMOLOGY
- INTERSTELLAR MATTER AND THE LOCAL UNIVERSE
- STARS AND STELLAR PHYSICS
- EDUCATION, OUTREACH AND HERITAGE
- HIGH-ENERGY PHENOMENA AND FUNDAMENTAL PHYSICS
- THE SUN AND THE HELIOSPHERE
- THE SOLAR SYSTEM, EXOPLANETS, AND ASTROBIOLOGY
- INSTRUMENTATION, SOFTWARE, LABORATORY ASTROPHYSICS AND DATA

Books in the program range in level from short introductory texts on fast-moving areas, graduate and upper-level undergraduate textbooks, research monographs and practical handbooks.

For a complete list of published and forthcoming titles, please visit iopscience.org/books/aas.

About the American Astronomical Society

The American Astronomical Society (aas.org), established 1899, is the major organization of professional astronomers in North America. The membership (~7,000) also includes physicists, mathematicians, geologists, engineers and others whose research interests lie within the broad spectrum of subjects now comprising the contemporary astronomical sciences. The mission of the Society is to enhance and share humanity's scientific understanding of the universe.

Creating the Molecules of Life

Richard N Boyd
The Ohio State University

Michael A Famiano
Western Michigan University

IOP Publishing, Bristol, UK

ISBN 978-0-7503-1993-5 (ebook)
ISBN 978-0-7503-1991-1 (print)
ISBN 978-0-7503-1992-8 (mobi)

DOI 10.1088/978-0-7503-1993-5

Version: 20181001

AAS–IOP Astronomy
ISSN 2514-3433 (online)
ISSN 2515-141X (print)

British Library Cataloguing-in-Publication Data: A catalogue record for this book is available from the British Library.

Published by IOP Publishing, wholly owned by The Institute of Physics, London

IOP Publishing, Temple Circus, Temple Way, Bristol, BS1 6HG, UK

US Office: IOP Publishing, Inc., 190 North Independence Mall West, Suite 601, Philadelphia, PA 19106, USA

We wish to dedicate this book to our wives, Sidnee and Nikki, for enduring the several months of our intense writing, editing, gathering permissions, and generally being extremely distracted.

Contents

Preface

Writing a book about the origin of life certainly requires some courage. The living creatures that exist now have very little in common with those that preceded them in the evolutionary paths that were taken. Even more seriously, there is very little hope of ever finding those that once existed on Earth, since the evolutionary winners tended to eat the losers.

Furthermore, the complexity of the chemistry that was necessary to get from simple molecules to the many forms of DNA that exist now in Earth's many living entities—small or large; simple or complex; animal, plant, or something else—is staggering.

And if that were not enough to stop anyone from pursuing such an endeavor, some of the steps necessary to explain even some of the better developed theories of the creation of life contain poorly understood steps themselves, creating gaps in theoretical or experimental knowledge. These continue to exist despite the incredible advances made in recent decades in quantum molecular chemistry and microbiology.

However, we have bitten off a small piece of the development of life, believing that much of what we suggest in this book is well enough understood that it makes sense to present its current state of the art. We will leave to others to work out the means by which the basic molecules of life—the amino acids and the nucleobases—developed into more complex molecules, and then into things that could satisfy some of the criteria for being alive. There are some suggestions as to how that came about, but the details remain to be discovered.

There is unassailable evidence, however, that those basic molecules were made in outer space, and that they came to Earth on meteorites. Furthermore, these amino acids that arrived from outer space have the property of handedness, and they apparently needed to have the correct handedness, that is, left-handedness, as an essential property for them to evolve into more complex structures. Indeed, they must have that same handedness to serve as the basis for feeding Earth's creatures of all types. Explanations of how they could have become left-handed after they arrived at Earth have found it difficult to overcome the challenges of those models.

There are several well-developed explanations of how amino acids achieved their handedness in outer space. We discuss several of these and try to evaluate their capability to describe how that handedness came about. We believe that the model that we have developed along with our collaborators comes as close as any in explaining how amino acids achieved their left-handedness in outer space. The explanation of our model involves extensive atomic physics and a little bit of nuclear and particle physics with some biochemistry thrown in for good measure, but explaining all of that is what most of this book is about.

We realize that this book may have an unusually diverse group of readers, including astrobiologists, astrophysicists, astronomers, physicists, and organic chemists. Because the overlap in knowledge bases between these groups is likely to be small, we have begun our discussions of each scientific discipline at a basic level. That means that most of the material in the first fews chapters will seem unnecessarily basic to some subset of the readers. We invite you to skip those sections.

Acknowledgements

We express extreme gratitude to our collaborators on the papers we have published on this subject: Toshitaka Kajino, Takashi Onaka, and more recently, Yirong Mo. All have made important contributions toward developing our model. We are also grateful to Laurence Barron and Michael Leyton, each of whom made several helpful suggestions.

R.N.B. also wants to acknowledge helpful interactions with Isao Tanihata and an extremely helpful discussion in 1989 with Reiko Kuroda.

Finally the coaching and encouragement we received from the people at IOP with whom we have interacted made it possible to produce this book. They include Ashley Gasque, Chris Benson, Dan Heatley, and Poppy Emerson.

Author biographies

Richard N. Boyd

Richard Boyd was awarded his PhD in nuclear physics from the University of Minnesota in 1967, having previously received BSE degrees in physics and mathematics from the University of Michigan. Following postdoctoral stints at Rutgers and Stanford Universities, he became an assistant professor at the University of Rochester, where he developed an interest in astrophysics. He then moved to The Ohio State University as an associate professor in physics, and was then promoted to full professor. While there, he also joined the Astronomy Department and served as an associate dean of the College of Mathematical and Physical Sciences for three years.

Boyd has made numerous visits to Japan, both to RIKEN, the Laboratory for Physical and Chemical Research, and to the National Astronomical Observatory of Japan. His interests in astrobiology began at RIKEN, developing slowly until, in 2010, he published his first paper in astrobiology with Toshitaka Kajino and Takashi Onaka.

Following his retirement from Ohio State, Boyd joined Lawrence Livermore National Laboratory (LLNL) as the science director for the National Ignition Facility for three years, retiring from LLNL following two more years in the Physics Division. He wrote a textbook, "An Introduction to Nuclear Astrophysics," which was published in 2007, as well as a book on astrobiology, "Stardust, Supernovae, and the Molecules of Life," published in 2012.

His hobbies include hiking, reading, writing books, and enjoying the natural beauty and agricultural products of Sonoma County, California.

Michael Famiano

Michael Famiano is an associate professor of nuclear astrophysics in the Department of Physics at Western Michigan University. After receiving his BS in physics at the University of Michigan, he received his PhD in nuclear physics from The Ohio State University. He then worked as a fellow for the Science and Technology Agency of Japan at the Institute of Physical and Chemical Research near Tokyo. He also held a postdoctoral position at the National Superconducting Cyclotron Laboratory in East Lansing, Michigan, before coming to Western Michigan University.

Famiano's primary research interest is in stellar nucleosynthesis. This includes the properties of heavy elements involved in stellar explosions and how these relate to the synthesis of the nuclei involved. He has worked as a visiting scientist at the National Astronomical Observatory of Japan and at Beihang University in Beijing to study this rapidly evolving field. His work involves theoretical modeling of astrophysical sites and experimental measurements of nuclear properties. He was awarded a Fulbright fellowship for his research in high-temperature and high-density plasmas in astrophysical environments.

Famiano currently resides in Portage, Michigan, with his wife, three children, and two dogs. He enjoys running, the outdoors, and encouraging his children not to grow up too rapidly.

Richard N Boyd and Michael A Famiano

Chapter 1

Introduction

1.1 In the Beginning

Our universe began 13.8 billion years ago. A short time later, at least by cosmic standards, stars formed and began to synthesize the elements that now populate Earth, our Galaxy, and much of the rest of the universe. In many locations, the molecules from which life ultimately evolved began to form. The processes by which stars formed often created planets, one of which was our Earth. And some of the molecules that formed in outer space got to our planet, there to begin their task of creating living entities.

The age of the universe is incontestable. It has been determined by several incredibly sophisticated experiments. The values obtained by those experiments for the age of Earth are determined using completely independent techniques, and they agree fairly well (although not perfectly, and that raises interesting questions, perhaps about the details of the Big Bang). These experiments are so important that we want to describe some of their details. They will be discussed in Chapter 2.

How old is Earth? It was born 4.55 billion years ago. The elements that exist on Earth, all of the carbon, oxygen, silicon, iron—indeed, everything except hydrogen and helium—were produced in stars that lived and died prior to the formation of our solar system. Because the universe is so much older than Earth, those elements are thought to have been produced in several previous generations of stars.

Earth's age is also incontestable; there are several ways to determine it, and their results are consistent with each other. One technique establishes a lower limit on Earth's age of around 4 billion years by dating the oldest rocks found on Earth. Radiocarbon dating is the most common form of this technique, but the relevant ages here are in the billions of years, and the half-life of ^{14}C, the radioactive isotope of carbon, is much too short (5730 years) to allow a meaningful measurement of something billions of years old (the maximum age for carbon dating is about 100,000 years). However, determining long past ages using radioactive nuclei is a well-established technique; it has been applied to the determination of the age of the

doi:10.1088/978-0-7503-1993-5ch1

universe (Cowan et al. 1991, 1997), although newer techniques, to be discussed in Chapter 2, provide a much more precise answer.

For measurements of the age of Earth, nature has provided us with a special crystal: zirconium silicate, $ZrSiO_4$. Zircons, when they crystallize from magma, have been found to contain uranium (U), but absolutely no lead (Pb). However, the U isotopes eventually decay to Pb isotopes. ^{204}Pb (82 protons, 122 neutrons) is a decay product of ^{238}U (92 protons, 146 neutrons), and ^{205}Pb (82 protons, 123 neutrons) is a decay product of ^{235}U (92 protons, 143 neutrons). Thus, zircons provide a very direct way of measuring Earth's age, by measuring the abundances of the lead isotopes in zircons that also contain uranium. The two U isotopes have extremely long half-lives (0.704 billion years for ^{235}U; 4.468 billion years for ^{238}U), and they are known to high accuracy. As ^{235}U and ^{238}U decay, the amounts of ^{204}Pb and ^{205}Pb in the sample increase. So, by measuring the amounts of ^{204}Pb and ^{205}Pb and comparing those to the amounts of ^{235}U and ^{238}U in a zircon, one can determine its age. The oldest rock found appears to have sufficient zircon inclusions that suggest it is more than 4 billion years old.

How sure can we be that radioisotope dating gives correct answers? The results only depend on the half-lives and the accuracy of the abundance measurements. Many tests have been conducted to see to what extent extreme conditions, e.g., the high temperatures encountered in early Earth, could influence these half-lives. The answer is that such influence is well below the smallest uncertainties in the half-lives. Of course, the dating approach can only tell the ages of the rocks; it therefore only provides a lower limit on the age of Earth. There are other ways to determine the age of Earth, but the value that one achieves with all of these techniques is 4.55 billion years (Stassen 2005).

Before leaving the dating story, we note that zircons can also provide an age for the life forms that are found in ancient rocks. Unfortunately, these fossils are found in sedimentary rock, and zircons are definitely not sedimentary rocks. However, some sedimentary rocks that contained fossils have also been found to contain zircons. If the zircons were included in the sedimentary rock at the same time the fossils were entombed therein, then the age would be easy to determine. However, the zircons may well have been added to the sedimentary rock only after the rock has existed for a while. Moorbath (2005) estimated that the fossils in these rocks are 3.67–3.70 billion years old. We have left out the details, but their ages are quite well established.

So what was Earth like initially? It was a pretty miserable place by any standard. It was very hot; had essentially no atmosphere, at least that living creatures could breathe, and very little water; and was being continually bombarded by meteorites. Those meteorites are important; they will play a central role in the creation of life on Earth. However, in the early history of Earth, they would have made our planet a pretty unpleasant place to be in. In fact, it would have been uninhabitable, as huge meteorites would have created enormous displacements of material, the clouds from which would surely have extinguished any life that might have survived the unbearable temperatures. How do we know this? The Moon was presumably subjected to the same bombardment as Earth, and its cratered surface tells us that this was the unfortunate situation on both early bodies. This situation has been studied by Sleep et al. (1989), among others.

However, this intense meteoritic bombardment ceased about 3.7 billion years ago, again determined from the evidence obtained from the Moon. Although an occasional meteorite still does strike Earth, their number has greatly decreased from that of the earlier period. Why did this change come about? It appears that Jupiter served as a cosmic janitor. Because of its large size, Jupiter exerts a huge gravitational pull on objects that come close to it, and that gravitational pull on the objects it encountered allowed it to sweep out or divert most of the debris that were originally floating about the inner solar system. This made it possible for any life that might arise on Earth after the cosmic cleanup was completed to flourish and evolve without too much concern for large-scale planetary destruction, although that can still occur from hits from large meteorites that were missed in the Jovian cleanup of the solar system. Incidentally, it apparently took Jupiter more than a billion years to remove the meteoroids in the inner solar system.

Of course, Earth had to cool, and it had to develop an atmosphere that contained some oxygen before life—or at least life as we know it—could begin. We know Earth also somehow increased its water content to its present level. It is generally believed that water is more or less essential to the existence of life, although that bias could lead us to miss recognizing some forms of life. We will return to this in the last chapter. The importance of liquid water is emphasized in the article by Rothschild & Mancinelli (2001). They note that where there is liquid water on Earth, virtually independent of all other conditions, there is life. "Other conditions" here can mean extremes of heat and cold, alkalinity, and a whole host of other possibilities. Of course, the most extreme conditions necessitate the existence of creatures that can withstand the extremes; not surprisingly, these are known as extremophiles. A few of them are introduced in the final chapter.

Plaxco & Gross (2006) devote a lot of space in their book to a discussion of the importance of water to life of any kind. That discussion is sufficiently important that we will repeat the most salient points. The human body is about 70% water, so there would seem to be a very strong prejudice in human life forms for that particular liquid. This is seen in virtually every biological process in which our cells and bodies engage; we must have an effective solvent for all of these to happen. Water is the best known solvent.

Furthermore, water has the rather unique property that when it freezes, it expands. Thus, when lakes freeze, the ice does not sink to the bottom, which would make it much easier for such lakes to freeze solid, probably eliminating all life forms that might have developed in association with lakes. When water freezes in cells, the ice thus formed tends to rupture the cells. We will return to this when we discuss extremophiles in the last chapter.

Water also has the unusual ability to absorb heat. Thus, it is an excellent moderator of climate change. And, of course, it remains liquid over a 100 °C (180 °F) temperature range, yet another unusual characteristic, and one that is very accommodating for life. Water also has an unusually large dielectric constant, which makes it possible to take many chemicals into solution in water, partially explaining why it is such a good solvent. This is an important property for the functioning of our cells. Earth's water may have arrived in the form of comets, which consist largely of ice, so if enough of them got to Earth, they would have fulfilled that function.

Water was needed on Earth for another reason. As mentioned above, Earth's atmosphere needed to contain oxygen for many life forms to exist. There was a lot of oxygen on Earth; it is after all the third-most abundant element in the universe. However, it was so chemically active that it was all trapped in molecules, such as water, H_2O, and carbon dioxide, CO_2, and it needs to be in its diatomic form, O_2, to be useful to life. This conversion was performed by cyanobacteria, very early life forms that were created on Earth, which used sunshine, water, and CO_2 to produce oxygen. This did not happen quickly by human time standards, but it was pretty rapid on a cosmic timescale.

But all of these sidestep the question of how life actually began on Earth. There are certainly a number of questions on how that occurred, many of which represent active areas of current research. There are also numerous books that deal with various aspects of that question. We will not repeat much of what the other books cover, but we will focus on some issues that may be central to the presence of amino acids on Earth. Amino acids are the building blocks of the proteins on which our lives depend, and so in turn are critical components for the existence of life on Earth. The issues we will focus on tie together some astrophysical phenomena that occur on macroscopic scales with the microphysics of molecules and even of the constituents of the nuclei of the atoms in those molecules—the very large with the very tiny.

1.2 What is "Life"?

This might seem like a question that should have an obvious answer, but that does not turn out to be the case. Indeed, scientists have endeavored to answer that question for decades, and their efforts have been preceded by others spanning centuries. An excellent paper on the definition of life was written by Luisi (1998), and we shall use many of his results. Charles Darwin, from his 19th century perspective, did give some thought to what was required for life to have occurred in the first place and came up with this observation: "The hypothesis of an originary [sic] arising of life from the inanimate matter … at least offer the advantages to explain natural things by natural pathways, thus avoiding to invoke miracles" (quoted from Luisi 1998, p. 615). In the early 20th century, Alexander Oparin studied this question, which culminated in his book "The Origin of Life" (Oparin 1957, p. 73). He attempted to refine the definition of life by specifying its requisite conditions, which he concluded necessarily involved six properties: (1) capability to exchange materials with the surrounding medium, (2) capability to grow, (3) capability for population growth (multiplication), (4) capability to self-reproduce, (5) capability to move, and (6) capability to be excited. He also added some additional properties, such as the existence of a membrane and the "interdependency with the milieu." Although there are many other suggestions of what constitutes life, this gives some indication of the necessary complexity of the attempts to define it (Luisi 1998).

NASA, the United States' National Aeronautics and Space Administration, has a huge investment of both money and time in searching for life, and so it has attempted to provide a definition of its own: "Life is a self-sustained chemical system capable of undergoing Darwinian evolution." As noted by Luisi (1998), this attempts to move the definition of life into the molecular level. It could evolve even further by relating the definition to ribonucleic acid, or RNA, but that would quickly exceed the scope of this book.

Finally, before we get to the criticisms of these definitions of life, we offer one more: that by Paul Davies in his book "The Fifth Miracle" (Davies 1999). We have restructured and condensed his requirements, but basically they include the following:

- Living things are autonomous, that is, they have self-determination. Living creatures can in some sense decide things for themselves. Davies compares a live bird, which can decide where it is going and what it is doing, with a dead bird, which clearly is incapable of either. However, the two birds are essentially identical, biologically and chemically, so there is some special feature that the live bird has that the dead one does not have. This statement does not deal with the reasons why one bird is alive and the other is dead; it only recognizes that they are somehow different.
- Living things reproduce, creating offspring that ultimately strongly resemble the parents. This is clearly a requirement for continuation. This also requires that living things have some mechanism for information transfer from one generation to the next.
- Living things metabolize, that is, they process chemicals, thereby bringing energy into their bodies. Living forms must use their food to produce the energy they need to perform the functions they must in order to continue to live and reproduce.
- Living things have complexity and organization, that is, life forms are composed of many atoms, and the different specialized groups of atoms and molecules that exist in complex beings must work together in ways that make all of the systems behave as they must to continue living.
- The complexity that characterizes life must be organized; the separate components must work together. Although the systems in the two birds discussed previously are essentially identical, they have ceased to work together in the dead bird. In the living bird, the lungs take in oxygen and insert it into the blood stream. The blood nourishes the brain, which directs its operations, but it cannot do so without the oxygen that the blood transports to it. So, any failure of one of those systems will result in a lack of organization, and death.
- Living things must be capable of developing, or must be able to change in response to demands of the environment. They must be able to evolve, that is, their reproduction must include mechanisms for evolution and responses to natural selection. This not only determines the ability of any species to change

> in response to the needs imposed on it by external factors, but it also allows for its improvement to survive constant competition with other members of its species and with those of other species. This is done by tiny changes in the DNA (deoxyribonucleic acid) of the species, i.e., mutations, most of which are destructive and so do not produce offspring that can survive, or by imperfections in reproduction. But a few of the changes will improve the ability of the species to survive, and those adaptations will permit their recipients eventually to replace their predecessors. This is how Darwinian evolution capitalizes on the mutations that are produced by, among other things, the cosmic rays that continually rain down on Earth. We are in a state of constant, albeit slow, mutation!

So how well do these definitions work? They have some obvious problems. For example, the NASA definition really defines life for a population, since reproduction often requires two members of the species, so no single entity of these species can be alive under that definition. Other exceptions are pretty easy to come up with. For example, mules are incapable of reproduction, but are arguably alive. And viruses really complicate the definitions, since they do pretty much fulfill the requirements for categorization as a life form, but are not included as such by everyone. An interesting discussion of this issue is given in the book by Peter Ward (2005). Thus, the bottom line is that it is not so easy to define what it means to be alive.

Since molecular biologists have gotten involved in searching for the origins of life, it is natural that modern definitions of life would take us to that realm. One recent edition of the journal *Astrobiology* (Deamer 2010) contained several papers that dealt with the definition of life. We will summarize and reference a sampler of what might be involved in a definition of life based on the microscopic.

Deamer (2010) bases his definition on what are generally agreed to be the basic molecules of life: the nucleic acids, DNA and RNA, and the proteins, all of which are referred to as biopolymers. He then notes the processes by which the biopolymers are synthesized, that is, by combining amino acids and nucleobases. His definition of life defines polymer synthesis as the fundamental process that must occur in living entities, using the capabilities of nucleic acids to store and transmit genetic information and those of enzymes to act as catalysts for metabolic processes. Deamer also includes reproduction as fundamental to his definition, but casts it in molecular terms. Finally, he notes that cells can evolve.

Some of the verbiage in Deamer's definition is daunting for non-microbiologists. What is most interesting, though, is that many of the properties he lists are not too different from those on Davies' list, but rather have taken them to a microbiological level.

For completeness, we include some basic definitions of life forms, specifically archaea, bacteria, and eukaryota. Archaea are single-celled microorganisms, and their single cell has no nucleus or many of the other features that human cells, which are eukaryota, enjoy. However, archaea do have genes and enzymes that catalyze some metabolic pathways, which give them some similarity, in that regard, to

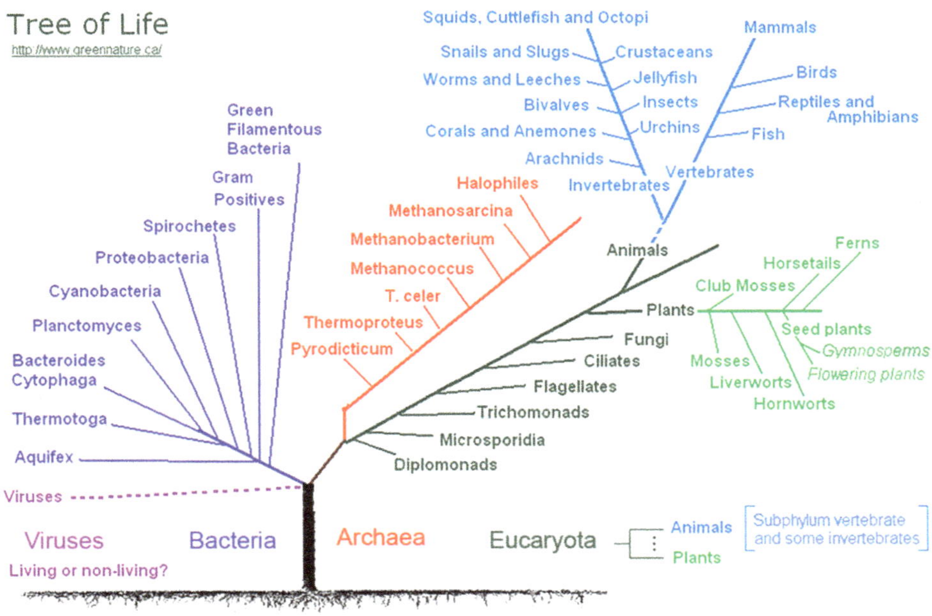

Figure 1.1. The Tree of Life, showing the three basic life forms we discuss, with many additional branches, and also the implication that there may be five or six kingdoms. Courtesy of greennature.com.

eukaryota, and they reproduce asexually. Bacteria are also single-celled microorganisms, and like archaea, their cells have no nucleus. Archaea and bacteria are roughly similar in size and shape, and often grouped together as prokaryota, that is, "not eukaryota." Prokaryota, simply stated, "are molecules surrounded by a membrane and a cell wall." The "Tree of Life," which provides some structure to the many possibilities, is illustrated in Figure 1.1.

Eukaryota are considerably more sophisticated than either archaea or bacteria, while having some similarities to both. Eukaryota and archaea are similar in that both have DNA, which governs their genetic machinery. Eukaryota differ from both archaea and bacteria, however, in that they have a nuclear envelope (Youngson 2006; Nelson & Cox 2005; Martin 1983), within which their genetic material is contained. They also have other cellular refinements that archaea and bacteria do not have, notably, many organelles, organs of the cell. All species of large complex organisms—animals, plants, and fungi—are eukaryota. Although human organisms are eukaryota, there are many more bacteria than eukaryota that inhabit the human body! Some of the bacteria are helpful, for example, in aiding digestion. However, some others can be dangerous; cholera and tuberculosis come to mind.

The first entity that could really be called living, that is, that would satisfy the criteria that define life, probably had only a few things in common with the modern definitions of archaea and bacteria, and even less with eukaryota. However, these first cells eventually evolved into the three distinct life-form branches of archaea, bacteria, and eukaryota. These are further divided into kingdoms, with archaea and bacteria comprising their own kingdoms, and eukaryota being divided into four: protista (unicellular protozoans and multicellular algae), plants, fungi, and animals. (But not everyone agrees on these kingdoms, or even that there are only six of them.)

Cell division in eukaryota is quite different from that in organisms that do not have nuclei. Although that is a very interesting subject, it is pretty far removed from our story, so we will leave that for other authors.

Fortunately, for the current book, we do not need to be too fussy about life's definition. As Justice Potter Stewart said in his famous quote about hard-core pornography, "I know it when I see it." Or will we? Some of the single-celled entities discussed in the previous paragraphs might not be so easy to recognize as living entities, especially after having been committed to the fossil record for a couple of billion years. So the answer to the "or will we?" question is not so obvious, and we will get back to it later. What we are presently concerned with is how the complex molecules that we know life requires are selected and propagated in outer space, which we will see that they are. Although the means by which they combine to form life are complex and poorly understood (and this is an incredibly important area of research), we will not worry about the details of how that comes about; it clearly does happen!

1.3 The Miller–Urey Experiment

Scientists have been trying to answer the question of our chemical origins for at least several decades. Alexander Oparin (1957, p. 73), mentioned above, and J.B.S. Haldane (1960) suggested that "a 'primeval soup' of organic molecules could be created in an oxygen-depleted atmosphere … through the action of sunlight. These would combine in ever-more complex fashions until they formed droplets," which would fuse with other droplets, and therefore reproduce via fission into daughter droplets. This would constitute a primitive metabolism in which those factors that are evolutionarily positive changes win out. Oparin and Haldane concluded that oxygen, which developed later on Earth, would prevent the synthesis of organic compounds that are necessary for life (although it is clearly essential for the continuation of many Earthly life forms, once they develop).

Half a century ago, Stanley Miller and Harold Urey (Miller 1953; Miller & Urey 1959), two chemists at the University of Chicago, performed an experiment to see if they could make amino acids from conditions that might have existed in early Earth. The idea for this experiment was Miller's, and he was a graduate student when he began this work. He had to convince his adviser, Urey, that they should do this experiment. However, when the work was published, Urey, already a very well-known chemist, was concerned that Miller would not receive appropriate credit for his work if Urey's name appeared on the paper. So, he had Miller publish the paper as the sole author.

There are hundreds of amino acids, but only 20 of them are the building blocks of the proteins that we require for life. Our bodies make about half of these, and we get the others from our food. Beyond those 20 amino acids, the rest are not essential for life, and most do not even occur naturally on Earth. Miller and Urey were trying to figure out how the first Earthly amino acids were produced. Their experiment was designed to see if the 20 on which we depend for life might have been created initially in an environment that probably existed early in Earth's history. What they found was that they could create at least some of the amino acids from a spark discharge in an oxygen-poor environment that contained a few basic molecules, consistent with Oparin's and Haldane's suggestion. If one assumes that such an environment could have existed on Earth in its early stages, one might conclude that the amino acids were formed on Earth in a lightning storm, and that subsequent chemical evolution produced all of the molecules that are essential to life.

Is that the end of the story? Not quite. Perhaps not at all. Amino acids also have a peculiar property, called chirality, that the Miller–Urey experiment does not explain. Chirality is another word for handedness. Your hands have chirality; if you hold them in front of you with the palms facing you, it is obvious that they do not look the same—the thumbs point in opposite directions. But if you stand by a mirror so that you are looking at the palm of one hand directly and at the palm of the other in the mirror, you find that now they do look the same. So, your hands do have a kind of symmetry—a mirror symmetry. From this you could conclude that there are two forms of chirality: left-handed and right-handed. It turns out that chirality also exists in many other molecules and in fundamental particles, and it is very much like it is with your hands; it is a mirror symmetry. For chiral molecules, you can think of each fingertip as containing a small group of atoms. Their mutual associations would lead to the same molecular binding energy, hence stability and most other properties, in either configuration. They would be essentially indistinguishable except for their chirality. Thus, the two forms have been given the designations left-handed and right-handed.

Chirality seems to manifest itself in a striking way in amino acids: it has been found that Earthly amino acids (with one exception, and it is achiral, meaning that it does not have two mirror symmetric states) are all left-handed! But if you produce them from a spark discharge in an early Earth chemical environment, you produce the same number of left-handed and right-handed ones. Therefore, there must be some other, or an additional, mechanism by which the amino acids are produced and processed that the Miller–Urey experiment does not address. As such, knowing how amino acid chirality came to be may well provide the key to understanding their origin.

It is now well established that amino acids are created in outer space and that they undergo some chiral selection there. The evidence for this is a result of meteorites that hit Earth, on which we will have more to say in a subsequent chapter. So, we are all extraterrestrials in a sense! We were obviously not born on another planet, but it does appear that the template molecules of the stuff we are made of were created far from Earth, and that the chirality of amino acids is a critical clue to their origin. It is clear that most of the elements that make up those molecules were created by stars and expelled into the vast reaches of the cosmos when those stars ended their lives. In addition, astronomers have found abundant evidence for complex molecules that have formed from those elements in outer space. Since some of those are the building blocks of the molecules of life, and because one of the mechanisms we will present in this book would establish the chirality that must be created in outer space, there is reason to believe that the basic requirements for life did indeed begin far from Earth, perhaps even before Earth existed.

That is what this book is about: it offers an explanation of how one aspect of life came to be the way it is. That falls under the rubric of what is known as panspermia: that our origins are in outer space. Panspermia also includes the idea that humans were conceived on another planet and then transported to Earth, but, given the difficulties and hazards of space travel, the scenario we present is perhaps more likely to be true. In any event, the origin of amino acids is a crucial key to understanding how mankind ended up on Earth.

1.4 General Background and Definitions

This book is about the theories that purport to explain how some of the seeds of life, the amino acids, were created in the cosmos and then transported to Earth to become a critical element in the creation of Earthly life. Many aspects of these theories are not new; scientists in ancient Greece discussed such concepts. Astronomical measurements have provided solid evidence that complex organic molecules are continually being created in the cosmos. And, as noted above, a handful of the meteorites that have crashed onto Earth's surface have demonstrated that amino acids and even more complex biologically relevant molecules are created in the cosmos, and can survive their trip to the surface of Earth.

However, as noted above, nature has given us a wrinkle that not only suggests that the molecules of life originated in the cosmos, but perhaps even *demands* that this be the case. This is the chirality of the amino acids. Subsequent chapters will present more details about chirality. Although the 20 amino acids on which our lives depend are all left-handed (except for the achiral glycine), there is no obvious reason why at least some of them could not have been right-handed, since experiments performed on Earth that make them have produced equal amounts of each. If we are to understand the origin of life, we must address this important clue to its creation.

Furthermore, the fact that the amino acids are all left-handed does not in and of itself require that they all have the same handedness, or even that they all have *some* handedness. However, this may well be the case, as Gol'danskii & Kuz'min (1989, pp. 1 and 6) state in their review article: "It can be concluded … that the biogenic scenario for the onset of the chiral purity of the biosphere could not, even in principle, have been realized in the course of evolution, *since without chiral purity of the medium the apparatus of self replication could not appear*. This apparatus is a basic process in the self-reproduction of any organism. *Life cannot arise in a racemic* [that is, equally abundant left- and right-handed] *medium*." (Italics are ours.)

Having defined racemic, we need to introduce two more terms: enantiomeric and homochiral. An enantiomeric medium has more molecules of one chirality than the other, but has some of each. If there is only one chirality, then the medium is homochiral. We define the enantiomeric excess, *ee*, as the relative difference between the number of left-handed and right-handed molecules in a population:

$$ee = \frac{N_L - N_D}{N_L + N_D},$$

where N_L and N_D are the number of left- and right-handed molecules, respectively. A positive (negative) *ee* indicates a larger number of left-handed (right-handed) molecules. A racemic mixture has an *ee* of zero.

The concept of chirality has existed since the time of Louis Pasteur (1948), who noted in 1860 that there is "a 'demarcation line' between life and non-life: the mirror dissymmetry of organisms" (quoted from Gol'danskii & Kuz'min 1989, pp. 1 and 6).

Trying to understand the origin of the chirality of amino acids and how that has influenced the development of life on Earth has challenged scientists for decades. In this book, we will concentrate on the first question and rely on subsequent scientific developments to deal with the second. Since this book is about beginnings, there are chapters on how the universe began, how the elements were produced, and why the cosmos favors left-handed molecules. This last item will require some details of core-collapse supernovae and other astrophysical sites, some information about neutrinos (particles that have no charge and almost no mass), and some basic nuclear physics. There are also explanations of how a small positive *ee*, once created, may need to be amplified in the cold confines of outer space, and then be amplified again when the molecules begin their planetary existence. That involves an explanation of the complex atomic physics that enables the selection of amino acids that are left-handed; they have left-handed chirality.

Although it seems improbable to involve the cosmos in determining something that seems as Earthbound as the properties of amino acids, consider the quote from a review article by Bonner (1991, p. 99): "The logical conclusion is that the source of terrestrial chirality must then have been extraterrestrial, and furthermore that it must have been capable of providing an ongoing influx of chiral molecules having uniform chirality."

Figure 1.2 gives you some indication of where we are going in general, and what our favorite theory for explaining amino acid chirality looks like. Any theory of amino acid creation that has them being produced in the cosmos will involve some of these same stages, so this figure contains more aspects of chirality selection than those of just our model. In our theory, the processing of amino acids to convert some of them to a preferred chirality occurs with an intense magnetic field and an enormous flux of neutrinos. These ingredients are naturally provided by supernovae, so we have called our model the Supernova Neutrino Amino Acid Processing (SNAAP) model. This particular figure is similar to those from works we did with our colleagues Toshitaka Kajino, Takashi Onaka, and Yirong Mo (Boyd et al. 2010, 2011, 2018; Famiano et al. 2018).

In our theory, the elements are mostly produced in stars, and that certainly is the case for the ones we are interested in, specifically carbon, nitrogen, and oxygen (excepting hydrogen, which was produced in the Big Bang). When a massive star completes all stages of its stellar evolution and explodes in a supernova, it seeds the

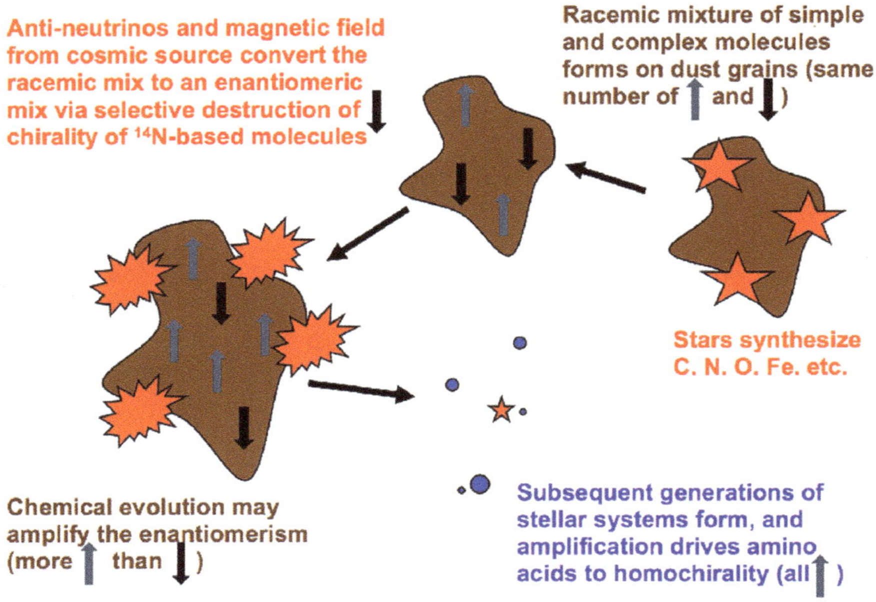

Figure 1.2. Scenario in which chiral amino acids are produced in the SNAAP model. The details are explained in the text.

cosmos with newly synthesized atomic nuclei. These elements are expelled in a huge cloud, called a nebula, which is created by the supernova explosion. As the nebula cools, atoms will form from the nuclei and electrons therein, then molecules will form on the surfaces of dust grains, or on meteoroids (meteoroids are the chunks of rock and possibly ice that exist as part of the stuff of outer space; when they pass through Earth's atmosphere and hit Earth's surface, they are designated meteorites). It is not that simple, but we will discuss this in subsequent chapters. Sometime later, a second star completes all stages of its stellar evolution and becomes a supernova, processing the molecules produced from the detritus of the first star through mechanisms we will discuss later. These mechanisms will certainly involve the antineutrinos emitted by some cosmic neutrino generator.

It may well be that not any supernova-producing star will do. Some stars that are sufficiently massive that they will ultimately become supernovae undergo a phase in their evolution called the red giant phase. This is accompanied by the expansion of the star's outer layer to about 1 astronomical unit, abbreviated au, which is the distance from our Sun to Earth. However, that will also turn out to be a bit more than the maximum distance for the supernova to produce a magnetic field large enough to be effective in producing amino acid chirality, and any amino acids within the processing distance will be inside the star. Hence, this ordinary supernova will not work.

However, a particular type of star, an especially massive one, that has become a Wolf–Rayet star prior to its explosion as a supernova may be able to supply the ingredients we need for our model (and also for some other models) to work while avoiding the red giant catastrophe. Wolf–Rayet stars expel their outer layer in an enormous wind so that it cannot become an extension of the star, thus eliminating their expansion in their red giant phase. Such stars will still become supernovae, possibly ultimately ending up as black holes, a particularly useful property to the SNAAP model (Boyd et al. 2010, 2018). When the Wolf–Rayet star explodes, it usually produces an intense magnetic field and always produces a huge neutrino flux. Some of the dust grains and meteoroids that were passing by will be close enough to the supernova that they, and the molecules they contain, will be affected by the intense neutrino flux and the magnetic field of the supernova. This combination, in interacting with the molecules, will produce a selective destruction of one orientation of the ^{14}N nuclei (an isotope of nitrogen, the nucleus of which is comprised of seven protons and seven neutrons) that exist in the amino acids that formed; this selects a particular chirality or handedness of those molecules. This chirality selection may possibly be amplified by the chemical replication that takes place on the molecules that have collected on dust grains that exist in the nebula and its surrounding region. Next, some of the enantiomeric-molecule-containing meteoroids will fall onto Earth and other planets, seeding them with molecules of that chirality. Finally, more amplification drives all of those molecules to homochirality, the condition in which only one chirality exists. This is what we have found to be the case on planet Earth.

There may be other possibilities for producing the intense magnetic fields and neutrino fluxes that are required to perform the chiral selection. One is a close binary star system in which one member is a neutron star and the other is a massive star. The outer shell of the massive star would be drawn into the gravitational potential well of the neutron star, forming an accretion disk around the neutron star. Planets from the massive star might end up circling both stars, but the mechanics of such a system are sufficiently complicated that a situation may exist in which a planet, complete with its amino acids, passes close to the neutron star, or is absorbed into the accretion disk. Atoms, molecules, dust grains, larger objects, planet-sized objects, and even amino acids can also form in the disk. The entire disk is typically close enough to the massive star that, when it becomes a supernova, its antineutrinos and magnetic field will process the entire disk and every amino acid that has formed in it.

As such, the idea that the chirality of amino acids arose in outer space, beginning with the production of the elements, followed by the formation of molecules on dust grain surfaces, and then by molecular amplification, is not new. What is new is the SNAAP model mechanism by which the molecules were processed into the chiral state that has been found to exist.

This is not the only model that has been invoked to explain amino acid enantiomerism. Several excellent books that attempt to describe how life began have been written. These include two books by Davies (1999, 2010) and one by Plaxco & Gross (2006). However, these books pay very little attention to the issue of amino acid chirality, which we believe to be central to understanding how and where the molecules of life were created. A book by Meierhenrich (2008) and review articles by Rode et al. (2007) and Avalos et al. (1998) give much more attention to the origin of chirality; we will return to some of the works of both the Meierhenrich and Rode groups later on. Another book, by Guijarro & Yus (2009), deals with molecular chirality as its primary subject. Also, as mentioned previously, articles by Bonner (1991, 2000) deal directly with the issues associated with amino acid chirality.

This book is not just about our theory. We will try to present samplers, and critiques, of some of the other theories that purport to explain how life as we know it, or in whatever form it might have taken, began.

References

Avalos, M., Babiano, R., Cintas, P., et al. 1998, ChRv, 98, 2391

Bonner, W. 1991, OLEB, 21, 59 Use of quote courtesy of Springer Nature

Bonner, W. 2000, Chirality, 12, 114

Boyd, R. N., Famiano, M. A., Onaka, T., Kajino, T. & Mo, Y. 2018, ApJ, 856, 26

Boyd, R. N., Kajino, T. & Onaka, T. 2010, AsBio, 10, 561

Boyd, R. N., Kajino, T. & Onaka, T. 2011, Int. J. Mol. Sci., 12, 3432

Cowan, J. J., McWilliam, A., Sneden, C. & Burris, D. L. 1997, ApJ, 480, 246

Cowan, J. J., Thielemann, F.-K. & Truran, J. W. 1991, ARA&A, 29, 447

Davies, P. 1999, The 5th Miracle: The Search for the Origin and the Meaning of Life (New York: Simon and Schuster)

Davies, P. 2010, The Eerie Silence: Renewing Our Search for Alien Intelligence (New York: Houghton, Miflin, Harcourt)

Deamer, D. 2010, AsBio, 10, 1001

Famiano, M. A., Boyd, R. N., Kajino, T., Onaka, T. & Mo, Y. 2018, AsBio, 18, 190

Gol'danskii, V. I. & Kuz'min, V. V. 1989, SvPhU, 32, 1, pps. 1 and 6 Use of quotes with permission of AIP Publishing

Guijarro, A. & Yus, M. 2009, The Origin of Chirality in the Molecules of Life (Cambridge: RSC Publishing)

Haldane, J. B. S. 1960, Natur, 185, 87

Luisi, P. L. 1998, OLEB, 28, 613 (Use of quote courtesy of Springer Publishing)

Martin, E. A. 1983, Macmillan Dictionary of Life Sciences, 2nd ed. (London: Macmillan)

Meierhenrich, U. J. 2008, Amino Acids in Chemistry, Life Sciences, and Biotechnology (Heidelberg: Springer)

Miller, S. L. 1953, Sci, 117, 528

Miller, S. L. & Urey, H. C. 1959, Sci, 130, 245

Moorbath, S. 2005, Natur, 434, 155

Nelson, D. L. & Cox, M. M. 2005, Lehninger Principles of Biochemistry, 4th ed. (New York: W.H. Freeman)

Oparin, A. I. 1957, The Origin of Life on Earth (tr. A. Synge; London: Oliver and Boyd)

Pasteur, L. 1948, Researches on Molecular Asymmetry of Natural Organic Products (Edinburgh: E. & S. Livingstone Ltd.)

Plaxco, K. W. & Gross, M. 2006, Astrobiology (Baltimore, MD: Johns Hopkins Univ. Press)

Rode, B. M., Fitz, D. & Jakschitz, T. 2007, Chem. Biodiversity, MD, 4, 2674

Rothschild, L. J. & Mancinelli, R. L. 2001, Natur, 409, 1092

Sleep, N. H., Zahnle, K. J., Kasting, J. F. & Morowitz, H. J. 1989, Natur, 342, 139

Stassen, C. 2005, The TalkOrigins Archive, www.talkorigins.org/faqs/faq-age-of-earth.html

Ward, P. 2005, Life as We Do Not Know It (New York: Penguin Books)

Youngson, R. M. 2006, Collins Dictionary of Human Biology (Glasgow: Harper-Collins)

AAS | IOP Astronomy

Creating the Molecules of Life

Richard N Boyd and Michael A Famiano

Chapter 2

What is the Origin of the Lightest Elements?

2.1 The Big Bang

Where do atoms come from? Most of the elements are made in stars. But things are not quite that simple. The elements hydrogen and helium comprise 99% of the ordinary matter of the universe, that is, the stuff that is not made of exotica such as dark matter or dark energy, and they were mostly produced in the Big Bang. The birth event of our universe is certainly the origin of everything we know, so let us begin our story with a discussion of the Big Bang. That name originated with Fred Hoyle, who actually believed in a steady-state universe, that is, one that did not have a birth event. Hoyle intended the name to be a pejorative comment on the model of his competitor. Of course, we now know that the name caught on, and the birth event of our universe is now a well-documented scientific paradigm.

Thirteen billion eight hundred million years ago, an extraordinary event occurred: our universe was born. This is well documented by many observations, but the Supernova Cosmology Project (SCP; Perlmutter et al. 1999), the High-Z Supernova Project (Riess et al. 1998), the Supernova HO for the Equation of State (SHOES) Collaboration (Riess et al. 2016), the *Wilkinson Microwave Anisotropy Probe* (*WMAP*; Bennett et al. 2003; Jarosik et al. 2011), and the *Planck* mission (Adam et al. 2016) stand out as modern incarnations of these efforts.

However, a very prominent forerunner of these efforts occurred in the early 20th century as a result of astronomical observations by Vesto Slipher and their interpretation by Edwin Hubble. Hubble was an interesting character, noted in his early life more for his athletic prowess than his academic abilities. He once won seven first places in a track meet, and he dabbled in amateur boxing for a time. He also got a law degree before serving in the military and then obtaining his PhD. Slipher was an astronomer. He had noted that the light from some galaxies appeared to be redshifted, that is, the characteristic wavelengths of the light from those galaxies could be identified as originating from emissions of photons—particles of light—from atoms of hydrogen, but they were shifted toward longer wavelengths.

(We will talk more about wavelengths in Chapter 4.) Since hydrogen is the most abundant element in the universe, it is appropriate to show some of the characteristic wavelengths that are emitted by hydrogen atoms; no other element emits light at those exact same wavelengths.

Figure 2.1 shows the different series (or groups) of emissions of photons when electrons in hydrogen atoms change from one allowed state to another. These are the results of transitions between specific energy levels that are allowed by microscopic physics, that is, the quantum mechanics of the particular atom. The Lyman series results from transitions from all higher lying energy states to the lowest lying level, called the ground state. The Balmer series is from transitions to the next highest energy level, called the first excited state. The Paschen series is to the second excited state. Each series has many lines, but they pile up at the leftmost line in each series. Those indicated in Figure 2.1 as being in the visible light region would be observed in the laboratory, that is, these are not redshifted. The Lyman series is shifted into the visible part of the spectrum in highly redshifted objects; these are the emission lines that were observed by Slipher.

Postponing the detailed discussion of wavelengths for a bit, we observe that light is electromagnetic radiation, and that it is characterized by an oscillating electric field and an oscillating magnetic field. The oscillations occur in both space and time. The wavelength is the distance over which a wave repeats itself. Visible light has a wavelength of around 5×10^{-7} m, or one-half of one-millionth of a meter. The above-mentioned characteristic wavelengths of the light from hydrogen in distant galaxies are Doppler-shifted because the galaxies are moving away from us. This is something that one experiences when hearing a train whistle or a police siren: the frequency of the sound is higher (which means that the wavelength is shorter) when the train or police car is moving toward us, then it drops as it passes by, because the wavelength becomes longer. The same effect applies to light. This ultimately led Hubble to make the incredibly bold assertion that all galaxies are moving away from all other galaxies in the universe, that is, that the universe was expanding in all directions no matter where one was in it. He also concluded from the amount of

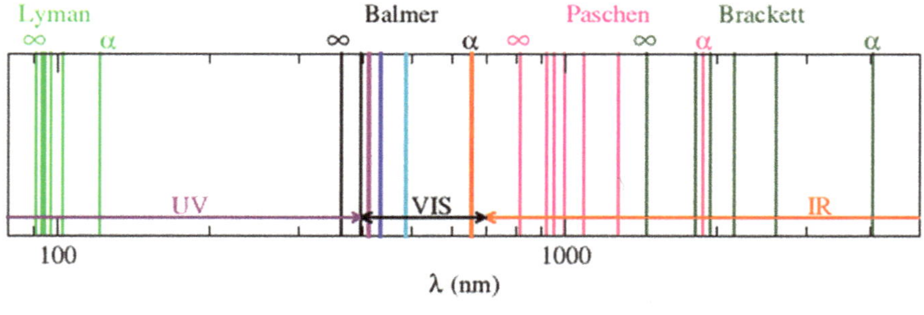

Figure 2.1. Characteristic emissions of hydrogen atoms. Those in the visible region, indicated by VIS, are shown by their color. Electrons in atoms can only exist at well-defined energies. Transitions between well-defined quantum mechanical states produce the observed photons, which are different for the different elements. Redrawn from a figure from ChemGuide.co.uk.

redshifts that the more distant the galaxies were, the faster they were receding. This led to Hubble's law:

$$v = HR,$$

where v is the velocity of recession between galaxies, R is their separation distance, and H is the constant of proportionality, called the Hubble constant. This law is a very simple-looking equation, but it has profound consequences: it says that the farther an object is from us, the more rapidly it will be receding from us, and this applies to every pair of objects in the universe! This law has prevailed for more than half a century, albeit with a large uncertainty on the value of the Hubble constant.

It is not that easy to envision what this universal expansion looks like in three dimensions, but if you can imagine our universe as being just two-dimensional, then you can think of it as existing only on the surface of a balloon. If you mark galaxies on the balloon then inflate it, you will see that every galaxy is receding from every other galaxy. In three dimensions, one can think of raisin bread as it is baking. As it rises, the distance between each pair of raisins increases, just as the distance between galaxies in our universe is constantly increasing—a direct result of an expanding universe.

Determination of the Hubble constant H led to a major irony of 20th century science. Two major groups of astronomers had been performing observations and analyses to determine H. One group, headed by the French astronomer Gérard de Vaucouleurs, consistently obtained values around 100 km s^{-1} Mpc^{-1} (a parsec is a unit of distance used by astronomers and is equivalent to 3.6 light-years, or 3.1×10^{13} [31 trillion] kilometers). The other group, headed by Alan Sandage, an American astronomer, consistently obtained values of around 50 km s^{-1} Mpc^{-1}, and the uncertainties quoted on their respective values were much smaller than the factor of 2 difference between them.

In science, when you have two results as discrepant as these, the last thing you would do is average them, since one of them is surely incorrect. Of course, both could be incorrect, and that turned out to be the case here. The modern value for H is about 70 km s^{-1} Mpc^{-1}, which is curiously close to the average of the Sandage and de Vaucouleurs results, and certainly the average of the two results within their uncertainties. However, it is a rough average of the most recent result from the SHOES Collaboration (Riess et al. 2016), which used the *Hubble Space Telescope* for its measurements, of 73.24 ± 1.74, and that from the *Planck* mission, 67.74 ± 0.46 (Adam et al. 2016). Might we be facing a "crisis in H" again but with much smaller error bars? We will comment more on this discrepancy below.

2.2 The Supernova Cosmology Project, the High-Z Supernova Project, and the SHOES Collaboration

However, the SC Project (Perlmutter et al. 1999) and the High-Z Project (Riess et al. 1998) both found, via very detailed measurements done in the 1990s, that the Hubble constant was not constant at all! Hubble's law had become so ingrained in astronomy that the redshift of distant objects was used to infer their distance from Earth. Thus, if one were to check Hubble's law, one would need some independent distance indicator. If you think about making an astronomical observation, you will quickly realize that it is easy to locate objects on an up–down and left–right plane, but determining the distance to an object, the third dimension, is much more difficult. What one needs is a class of objects that always exhibits the same intrinsic brightness, or which produces some other observable quantity that allows the determination of their intrinsic brightness. Then, the observed brightness will allow one to infer the distance to the object, since it falls off as the inverse square of the distance to the object. One class of objects used to infer distance is the Cepheid variables, stars for which their brightness oscillates with a frequency that can be related directly to the intrinsic brightness. Thus, astronomers can measure the frequency of oscillation of a Cepheid and determine its intrinsic brightness. Comparing that to the observed brightness then gives the distance to the star. Unfortunately, Cepheids are not especially bright stars, so some other standard candle was needed for making measurements at the huge distances that characterize cosmology.

Such objects are Type Ia supernovae (Hamuy et al. 1996; Boyd 2007). These are extremely bright exploding stars that are all essentially of the same mass before they explode. Therefore, since they explode by thermonuclear runaway and blow up the entire star, and the nuclear processes are essentially the same for all Type Ia supernovae, they have nearly identical intrinsic brightness. The SCP, High-Z Project, and newer SHOES Collaboration all utilized Type Ia supernovae as their large-distance standard candles.

The simplest description of the universe would be that it formed in a giant explosion, and has been expanding and slowing its rate of expansion ever since. The reduced rate of expansion would result from the gravitational attraction of all the constituents of the universe on each other. What the three projects found, however, was that although the universe is expanding, the expansion was speeding up, not slowing down. This suggests that there is something that acts like a negative gravity, that is, it does exactly the opposite of what gravity does. This has been dubbed dark energy. Adam Riess and Brian Schmidt, the leaders of the High-Z Project, and Saul Perlmutter, the leader of the SC Project, were awarded the 2011 Nobel Prize for this discovery. The existence of dark energy actually harkens back to Einstein, who included it in his general relativity equations and called it a cosmological constant. He later referred to it as his greatest mistake!

Nearly a century later, we have come to realize that, as has often been the case, Einstein was ahead of his time and was way ahead of his fellow scientists. However, his cosmological constant did not arise because of any data, but just as a result of his mathematical derivations. In any event, determining what dark energy is and understanding why it acts the way it does will constitute one of the primary objectives of scientists for at least the next decade.

The results from the SHOES Collaboration are shown in Figure 2.2, which plots the distance ladders, the bootstrapping of the nearest distance indicators, those determined from the parallax generated by the extremes in Earth's orbits, with the intermediate-range ones, the Cepheid variables, and those that can indicate greater distances, the Type 1a supernovae, used in their determination of the Hubble constant. Only recently has it become possible to obtain distance indicators that avoided the use of redshifts, which had been the historical distance indicator. Since the light from more distant objects has been traveling longer than that from closer objects, it has a greater lookback time, to use astronomers' jargon. More distant objects still recede at higher velocities than closer ones, but the higher precision of modern measurements makes it clear that recession is no longer described by an equation as simple as Hubble's Law. The data in Figure 2.2 appear to lie along a straight line, if one does not look too closely, although the fit to them is far more

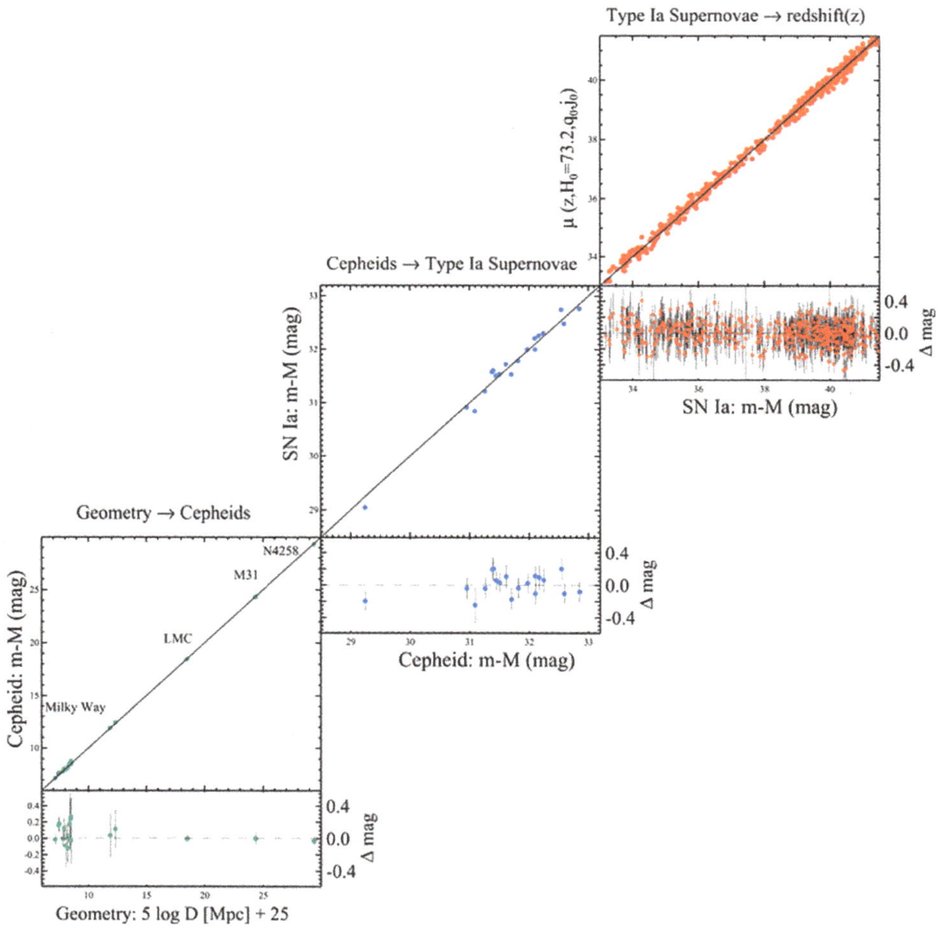

Figure 2.2. Complete distance ladder used by the SHOES Collaboration in determining the Hubble constant. The simultaneous agreement of pairs of geometric and Cepheid-based distances (lower left), Cepheid- and SN Ia-based distances (middle panel), and SN- and redshift-based distances provides the measurement of the Hubble constant. For each step, geometric or calibrated distances on the *x*-axis serve to calibrate a relative distance indicator on the *y*-axis. Results shown are an approximation to the global fit discussed in Adam et al. (2016). Reproduced from Riess et al. 2016, "A 2.4% Determination of the Local Value Of The Hubble Constant," *ApJ*, 826, 56. Courtesy of A. G. Riess.

complicated than that and involves a set of cosmological parameters, only one of which is the Hubble constant.

The data do not favor a scenario in which the universe is expanding at a decreasing rate as a result of only mutual gravitational attraction. Rather, they prefer a scenario in which the universe is not only expanding, but is doing so at an accelerating rate. Hence, the thing to take away from this is that the results of the SCP (Perlmutter et al. 1999), the High-Z Project (Riess et al. 1998), and the SHOES Collaboration (Riess et al. 2016) do not support the model that had prevailed for more than half a century, but do support a considerably more complicated universe—one that contains stuff that scientists, excepting Einstein (with his cosmological constant), had not imagined prior to the recent astronomical results.

From the SHOES Collaboration data shown in Figure 2.2, one can see the attention that was given to using all possible distance indicators, making sure they agreed in regions where they actually overlapped.

2.3 The *Wilkinson Microwave Anisotropy Probe*

Although the SCP, High-Z Project, and SHOES Collaboration used the most direct way to check the veracity of Hubble's law, these were not the only experiments done to determine cosmological parameters. As mentioned above, the *WMAP* and *Planck* cosmological mission are examples of another class of projects to study cosmology; they measured fluctuations in the 2.7 K (this corresponds to −270.5 °C, 4.9 °R, and −454.8 °F; this is pretty cold on any temperature scale) cosmic microwave background radiation. This is the electromagnetic radiation—the photons—left over from the Big Bang. They were first discovered by accident by Arno Penzias and Robert Wilson, two scientists at Bell Laboratories in New Jersey, as they were trying to develop an extremely sensitive antenna.

Penzias and Wilson were unable to eliminate some background noise, despite heroic efforts to do so, including chasing out some pigeons who had nested in their antenna. Fortunately, a few miles down the road at Princeton University was Robert Dicke, a theoretical cosmologist, who explained to Penzias and Wilson that they would never get rid of the background noise, since it was a relic of the radiation produced in the Big Bang. This radiation was very hot at the time of the Big Bang, but as the universe expanded, the wavelength of the radiation lengthened with the expansion. Longer wavelengths mean less energetic radiation, so this radiation is now extremely low energy. Penzias and Wilson won the Nobel Prize for their discovery. A much more elegant experiment was performed in the late 20th century by a team led by George Smoot and John Mather (Smoot et al. 1992); they measured the 2.7 K cosmic microwave background radiation in much greater detail, for which they were also awarded the Nobel Prize.

However, the Smoot–Mather measurement underwent an incredibly sophisticated improvement with the *WMAP*, which was designed to measure temperature fluctuations in the background radiation rather than the temperature itself. The *Planck* detector improved even more on the *WMAP* results.

The density in the early universe was blotchy, and the sizes and densities of the blotches tell a great deal about the status of the universe at the time electrons were captured on nuclei to form neutral atoms. Prior to this, electrons had been free because the temperature of the universe had been too high—and therefore the density of photons with sufficient energy to ionize the atoms was too great—for atoms to exist. The measurement of the blotches can therefore provide tests of theories of how the early universe formed and evolved. *WMAP*'s sensitivity was such that it was able to measure the fluctuations to a few parts in a million.

The results from *WMAP* (Bennett et al. 2003; Jarosik et al. 2011) showed that the baryon density of the universe—baryons are the protons and neutrons that comprise the nuclei of the atoms of which we are made—constitute less than 5% of the mass energy of the universe. This should produce some level of humility. Copernicus realized that the Earth was not the center of the solar system, but rather that it orbited around the Sun, which he concluded was the center of the universe. In 1750, Thomas Wright correctly surmised that the Milky Way might actually be a collection of a huge number of stars held together by their mutual gravitational forces.

In 1917, Heber Doust Curtis observed that novae in the Andromeda "Nebula," as it was then called, were drastically fainter than novae in the Milky Way. Curtis concluded that Andromeda was much farther away than had been previously assumed. He concluded that a number of objects previously believed to lie within the Milky Way were actually independent galaxies.

In 1920, the Great Debate was staged between Curtis and another well-known astronomer, Harlow Shapley. The topic, described by the title "The Scale of the Universe," concerned several aspects of astronomy. Curtis argued that our Galaxy was just one of many such collections of stars, while Shapley maintained that it was the only such collection, and the groups of stars that Curtis was placing at huge distances actually resided in our Galaxy.

Edwin Hubble settled the debate about the existence of other galaxies in 1925 when he identified and used extragalactic Cepheid variable stars for the first time to estimate extragalactic distances. His measurements demonstrated conclusively that the Andromeda Nebula was not a cluster of stars and gas within our Galaxy, but an entirely separate galaxy located a large distance from our own. This proved unquestionably the existence of other galaxies.

Thus, not only are we not the center of the universe or even of the Galaxy, we are less than 5% of the stuff from which the universe is made!

2.4 The *Planck* Cosmological Mission

In 2009, the European Space Agency launched a mission, *Planck* (Adam et al. 2016), named after Max Planck, the discoverer of blackbody radiation, that was also designed to map the cosmic microwave background, although with a considerably higher precision than *WMAP* could provide. *Planck* had two detectors, both cooled to near absolute zero in order to observe the low-frequency radiation they were

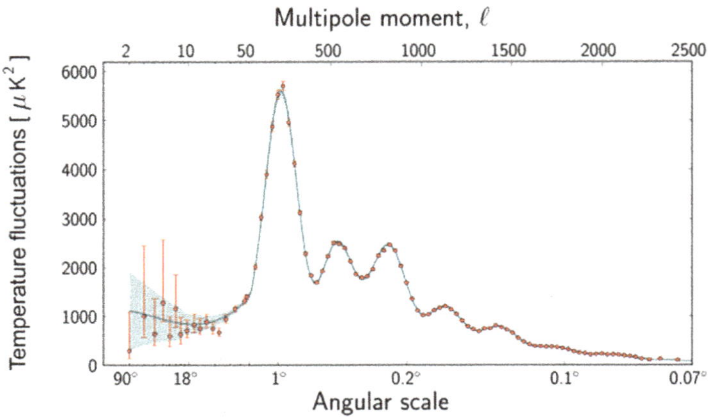

Figure 2.3. The *Planck* 2015 temperature power spectrum. This graph shows the fluctuations in the cosmic microwave background detected by *Planck* at different angular scales (or, alternatively, multipole moments) on the sky, starting at 90° on the left side of the graph, through to the smallest scales on the right-hand side. The blue curve shows the theoretical fit to the data with the standard cosmological model. Based on observations obtained with *Planck* (http://www.esa.int/Planck), an ESA science mission with instruments and contributions directly funded by ESA member states, NASA, and Canada. Courtesy of ESA, NASA, and ESA Canada.

looking for. These operated in nine wavelength bands (compared to five for *WMAP*), ranging from microwave to the very far infrared.

The results are shown in Figure 2.3 with error bars (where they are larger than the dots), which account for the measurement errors as well as for an estimate of the uncertainty due to the limited number of points in the sky where it is possible to perform each measurement. This figure shows the fluctuation in temperature over an angular range specified on the *x*-axis. For a personal calibration on the meaning of the angular sizes, the diameter of the full Moon is about half a degree.

The curve represents the best fit of the standard model of cosmology—currently the most widely accepted scenario for the origin and evolution of the universe—to the *Planck* data. The pale green area around the curve shows the predictions of all the variations of the standard model that best agree with the data.

Both the *Planck* and *WMAP* data showed that 23% of the universe is dark matter. This is stuff that interacts very weakly with most probes that one might devise to look for it, but it does produce a gravitational effect in galaxies and so is obviously present from the motions of galactic constituents. The third peak from the left is especially sensitive to the amount of dark matter. Finally, the dominant component of the universe's mass energy, 72% of it, is dark energy, as determined from both the SCP, High-Z, and SHOES results and the *WMAP* and *Planck* results. The two types of experiments also determined that the age of the universe, to high accuracy, is 13.8 billion years.

Interestingly, as noted above, the precision in the value obtained for the Hubble constant from the *Planck* mission data is significantly lower than that obtained from the SC and High-Z Project determinations and from more recent SHOES measurements using the *Hubble Space Telescope* (Riess et al. 2016). This may reflect the times following the Big Bang that are relevant to the two types of measurements. The *Planck* data determine the cosmological parameters as they existed 380,000 years after the Big Bang, whereas those from measurements relying on Type IA supernovae as standard candles stretch back as far in time as possible from the present. At the time of writing, astrophysicists are trying to figure out if this indicates new physics, a misinterpretation of data, or is just a stretch of the statistical uncertainties.

At present, scientists know what baryonic matter is, but do not know what comprises dark matter, and we do not have any idea what dark energy is!

2.5 Olbers' Paradox

There is an interesting argument that shows that the steady-state universe, Fred Hoyle's favorite cosmological theory (Hoyle et al. 1993), cannot possibly be correct, or, at least in its simplest form, has serious problems. This is known as Olbers' Paradox. Simply stated, this asks why the night sky is dark with speckles of starry light instead of being bright. Olbers' Paradox (Wesson 1991) was promoted in the 19th century by astronomer Heinrich Olbers, although the idea behind it was apparently realized as early as the 16th century by Kepler.

Why might the night sky be bright? Consider Figure 2.4, which shows a star seen by an observer, who is at distance 1 away from it. Suppose now that the observer is at distance 2 from another section of the sky. That volume of space is twice as far from the observer as that at distance 1, and so contains four times as many stars as that at distance 1. Assuming all of the stars are equally bright, each will seem one-fourth as bright to the observer as the star at distance 1. If the observer now looks at stars in a shell that is three times as distant as that at distance 1, each star will be one-ninth as bright as the star at distance 1, but there will be nine times as many of them, and so forth. Each shell thus contributes the same amount of light to the observer!

Thus, if you keep adding shells out to infinity, you should just keep adding to the light seen by the observer, and since each shell gives the same amount of light, things will be bright indeed! Note that it was assumed that the universe was both uniformly

Figure 2.4. The observer in relation to different sections of the sky. Section 2 (3) is at twice (thrice) the distance as Section 1, and so contains four (nine) times as many stars.

populated with equally bright stars and that it is infinite. These are assumptions that are basic to the steady-state universe theory.

But because the night sky is not bright, there must be something wrong with the assumptions that went into Olbers' Paradox. First, we know that the universe is not infinite, although it is pretty huge, but that does not completely resolve the problem. Second, stars are not constant emitters of light; they evolve. Third, the universe is not static; it is not in a steady state. We know that it is expanding, and this will increase the wavelengths of the radiation from the distant stars, which will also decrease the energy of the photons. In fact, the energies of the light from sufficiently distant stars will be so low as to be unobservable to the observer's eyes. Furthermore, the universe is, in a sense, young, in that the light from distant stars has not had time to reach us. This certainly would not be the case if we had a steady-state universe (Bahcall et al. 2005). Hence, Olbers' Paradox is not really a paradox at all when viewed from the perspective of modern cosmology.

2.6 Big Bang Nucleosynthesis

To describe what nuclides are synthesized in the Big Bang (Beringer et al. 2012), we need to go through the nuclear reactions that produced the nuclei. These will rely on some physics conservation laws, which we will explain as we proceed. Actually, very few nuclei were synthesized, due to a couple of nuclear quirks: there are no stable mass 5 or mass 8 nuclides. Mass 5 would be ^{5}He or ^{5}Li, and mass 8 would be ^{8}Be. To define the nomenclature, the superscript indicates the total number of protons and neutrons in the nucleus, with the number of protons being indicated by the element name. He is a nucleus with two protons and Li has three, while the total number of protons + neutrons is five in either nucleus. ^{8}Be (with four protons and four neutrons) is close to being stable, but it only lives 10^{-16} (one ten-millionth of one-billionth) of a second. The two mass 5 nuclei are much too unstable to live for even a short time. These facts make it virtually impossible to form anything by Big Bang nucleosynthesis (BBN) except ^{2}H, ^{3}He, ^{4}He, and ^{7}Li; in fact, ^{7}Li is so difficult to produce that its abundance is extremely small.

Seconds after the Big Bang, the only particles were protons and neutrons (and electrons). As the universe was expanding, it was also cooling, but it needed to cool quite a bit before the very first reaction could take place. That reaction is

$$^{1}\mathrm{H} + \mathrm{n} \rightarrow {}^{2}\mathrm{H} + \gamma,$$

where ^{1}H and n refer to protons and neutrons, ^{2}H is a heavy hydrogen nucleus—a deuteron, comprised of a proton and a neutron—and γ is a gamma-ray, a very energetic particle of electromagnetic energy. All particles of electromagnetic energy are called photons. A photon is a necessary component of that reaction in order for it to conserve energy. This includes not only the energy of motion, but also mass energy, that is, $E = mc^2$, Einstein's famous equation. Thus, energy conservation says that the sum of the mass energies and the energies of motion of the particles on the left-hand side of the equation must equal those same quantities on the right-hand side.

There are two more conservation laws that we need to attend to in this and all the other reactions that follow. The first is charge conservation. In the above equation, the proton has a charge of +1 in units of electronic charge, and the neutron has zero charge. On the right-hand side, the deuteron also has a charge of +1, and the gamma-ray has zero charge. So, each side has a charge of +1, and charge is conserved. If there had been an electron in the equation, its charge would have counted as −1. The other law that must be satisfied is baryon conservation. For present purposes, baryons are just protons and neutrons and nothing else (there are others, but they occur at much higher energies than our discussion of BBN will be dealing with), so the number of protons and neutrons on the left side of the equation, including those that exist in nuclei that contain both protons and neutrons, has to be equal to that number on the right side. (By the way, baryon conservation is thought by particle physicists to be violated, but only at an incredibly tiny level that does not affect BBN.) Gamma-rays are not baryons, so there is no conservation law that affects them, aside from the conservation of energy. The deuteron is an unusually loosely bound nucleus. Generally, it takes around 8 MeV—million electron volts—a unit of energy that is appropriate to nuclei, to liberate a single proton or neutron from a nucleus, but the deuteron can be broken into its two constituents with only 2.2 MeV, a pretty small amount of energy by nuclear standards.

So, the deuteron really was the bottleneck that required the universe to cool before BBN could begin. However, once ^{2}H began to be formed, BBN began in earnest, as can be seen in Figure 2.5. There was also a bit of a contest going on. A free neutron is not a stable particle; it decays to a proton, an electron, and a neutrino (technically, an electron antineutrino) with a half-life of just a little over 10 minutes. However, most of the neutrons would have been captured into nuclei before they

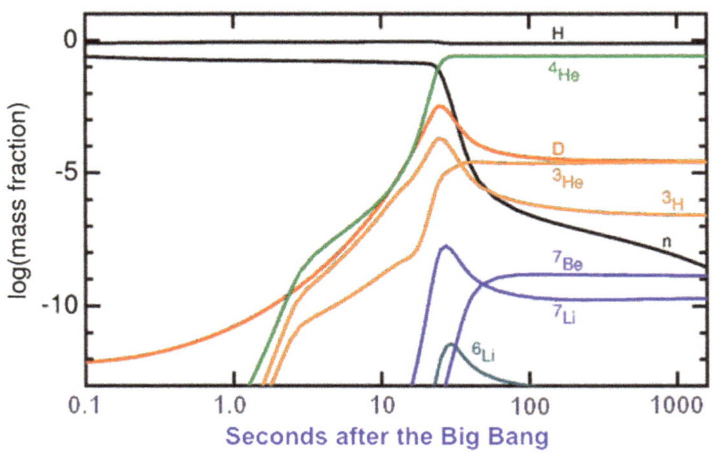

Figure 2.5. Evolution of BBN abundances during the time they are being produced, seconds after the Big Bang. It can be seen that the deuterium abundance (D) must increase before anything else can, but that the ^{4}He abundance increases rapidly as soon as D begins to form. From Burles et al. 1999, arxiv:astro-ph/9903300. Courtesy of M. Turner.

have a chance to decay, and they are stable in their nuclear homes, provided that the resulting nucleus is stable.

The reactions that convert protons and neutrons into ^{4}He nuclei—comprising two protons and two neutrons—are as follows:

$$^{2}H + {}^{1}H \rightarrow {}^{3}He + \gamma,$$
$$^{3}He + n \rightarrow {}^{4}He + \gamma \text{ or } {}^{3}He + {}^{2}H \rightarrow {}^{4}He + {}^{1}H,$$
$$^{2}H + n \rightarrow {}^{3}H + \gamma,$$
$$^{3}H + {}^{1}H \rightarrow {}^{4}He + \gamma \text{ or } {}^{3}H + {}^{2}H \rightarrow {}^{4}He + n.$$

What is going on in these reactions is as follows. In the first of the two pairs of equations, a proton is captured onto a deuteron, making ^{3}He (an isotope of helium that has two protons and one neutron); this is then converted to ^{4}He, either with a neutron capture or in a reaction wherein a deuteron adds its neutron to ^{3}He and releases its proton. In the second pair of reactions, a neutron is captured onto a deuteron to make ^{3}H, a triton (an even heavier isotope of hydrogen than the deuteron, since the triton has a proton and two neutrons), which then gets converted to ^{4}He either by capturing a proton or in a reaction wherein a deuteron drops off its proton and liberates its neutron. Note that each of these reactions conserves baryons.

That is pretty much all there is to BBN, except that ^{7}Li (with three protons and four neutrons) can be made in tiny amounts by either the first reaction below or the reaction followed by a decay, where e^- is an electron and ν_e is an electron neutrino:

$$^{4}He + {}^{3}H \rightarrow {}^{7}Li + \gamma, \text{ or}$$
$$^{4}He + {}^{3}He \rightarrow {}^{7}Be + \gamma,$$

and then

$$^{7}Be + e^{-} \rightarrow {}^{7}Li + \nu_e.$$

On the last line is the reaction by which ^{7}Be, which is not a stable nucleus (it consists of four protons and three neutrons), ultimately decays to ^{7}Li by capturing an electron, as indicated. Neutrinos are very important to our story, but not so much for BBN. However, for the moment it is worth noting that our Sun emits 1.8×10^{38} (100 trillion trillion trillion) neutrinos per second, 8.4×10^{28} (10 thousand trillion trillion) of which impinge on one side of the Earth, and virtually all of which pass right on through. If you close your fist, you will be enclosing a volume that contains several hundred neutrinos. Of course, it is not the same several hundred neutrinos for very long; the neutrinos will pass through your hand with virtually no recognition of your presence. But they are moving fast; 100 billion of them pass through each of your thumbnails every second.

Solar neutrinos were the source of one of the major scientific puzzles of the 20th century, that is, why was the rate of detection of solar neutrinos about one-third of that predicted by the standard solar model (see the website of the late John Bahcall, http://www.sns.ias.edu/~jnb/, for many discussions of the solar neutrino problem and the standard solar model, and Bahcall et al. 2005). The solution to this puzzle required the efforts of many physicists for several decades and uncovered a profound aspect of neutrinos: they can change from one type—called flavor—to another. We will have more to say about neutrinos later. Finally, note that neutrinos and electrons are not baryons. They are of another class of particles called leptons, and they must also be conserved in the reactions relevant to BBN and our considerations in subsequent chapters. Note, though, that if a lepton and an antilepton exist on one side of an equation, the net number of leptons is zero: for purposes of lepton conservation, the lepton cancels the antilepton.

The abundances of the nuclides made in BBN (relative to that of hydrogen) and the predicted abundances are shown in Figure 2.6, where they are plotted as a function of the baryon-to-photon ratio, which is essentially the baryonic density fraction of the universe. Prior to *WMAP*, that ratio was not well known, so it was customary to plot the BBN abundances as a function of that density as a way of determining its value. The vertical region labeled CMB in Figure 2.6 gives the *WMAP* value, which is seen to concur with the shaded strip labeled BBN, the preferred region determined from the BBN abundances, which were determined from the ^{2}H and ^{4}He abundances. The shaded CMB area falls within the preferred ^{4}He region, if one includes all of the uncertainties, both statistical and systematic (2σ means that the boxes extend to the 95.4% confidence level, that is, there is only a 4.6% chance that the value of any measurement will lie outside the indicated error boxes). Systematic errors are especially important in determining the primordial abundance of ^{4}He.

The agreement with ^{4}He is important because nearly all of the neutrons that existed in the early universe are predicted to end up in ^{4}He nuclei, so its BBN value represents the result of a simple prediction. However, astronomers are confident that they have observed ^{2}H in environments that come close to representing the Big Bang abundances, so its value is thought to reflect the Big Bang value. Also, its predicted value is in excellent agreement with the observed value at precisely the *WMAP* baryon density. ^{3}He is both made and destroyed in stars, so its BBN value has a greater uncertainty than that of the other BBN nuclides; therefore, its BBN value is not usually included as part of the success/failure criteria of BBN theory.

What about ^{7}Li? At low baryon density, it is mostly made by the $^{3}H + {}^{4}He \rightarrow {}^{7}Li + \gamma$ reaction, whereas at higher baryonic density, mass 7 nuclei are mostly made by the $^{3}He + {}^{4}He \rightarrow {}^{7}Be + \gamma$ reaction, and ^{7}Be subsequently captures an electron to make ^{7}Li, emitting an electron neutrino in the process (and therefore conserving leptons). The mass 7 nuclide production from the former reaction falls off as the baryonic density increases before the latter reaction fully takes over, which is what produces the dip in the ^{7}Li abundance curve. The bottom of the dip is just about what is observed for the Big Bang ^{7}Li abundance. Unfortunately, the CMB value is at a higher baryonic density, and the ^{7}Li abundance at that value is about a factor of 3 above the observed value. The resolution of this discrepancy has been the subject of an enormous amount of research; this is an ongoing topic for many cosmologists and astronomers.

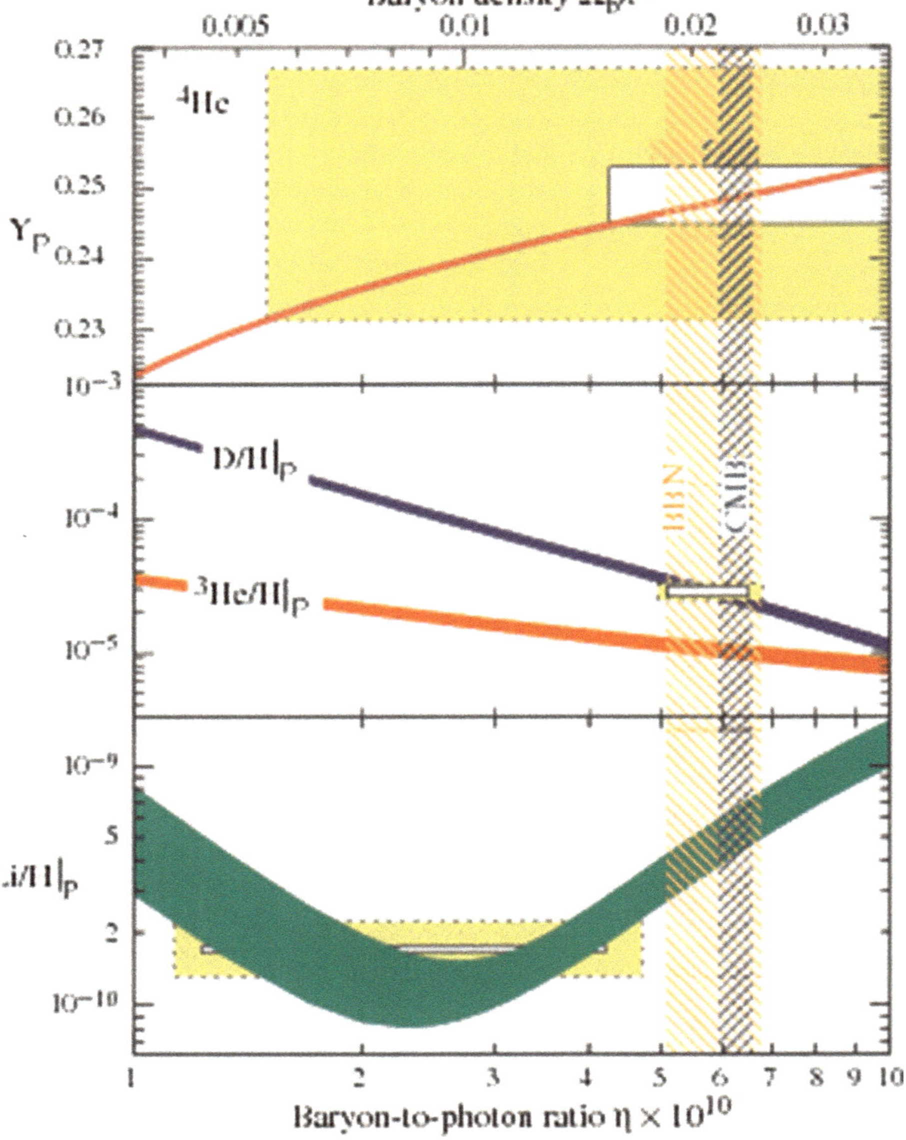

Figure 2.6. The abundances of ^{4}He, D, ^{3}He, and ^{7}Li as predicted by the standard model of Big Bang nucleosynthesis—the bands show the 95% confidence limit range. The boxes indicate the observed light element abundances. The narrow vertical band indicates the CMB measure of the cosmic baryon density, while the wider band indicates the BBN concordance range (both at 95% CL). Image courtesy of Brian D. Fields. Reprinted figure with permission from J. Beringer et al. (Particle Data Group), "Review of Particle Physics," Phys. Rev. D, 2012, 86, 010001. Copyright 2012 by the American Physical Society.

So how is one to solve the lithium problem? Recent studies (Boyd et al. 2010; Chakraborty et al. 2011; Cyburt & Pospelov 2012) have looked at possible nuclear reaction solutions, that is, reactions that might contribute to BBN, but which are not well characterized in the BBN computer codes. One particularly interesting aspect of these studies is that, since ^{7}Be is the source of most of the ultimate ^{7}Li abundance during BBN, reactions that might destroy ^{7}Be would mitigate the discrepancy between observation and theory. (Incidentally, there are potentially a lot more reactions that could occur than are indicated above!) All of the above reaction studies identified the ^{7}Be + ^{2}H reaction as the most promising candidate for a nuclear physics solution, although one of the studies (Cyburt & Pospelov 2012) argued that it could not contribute what would be needed to reduce the ultimate ^{7}Li abundance by the required factor of 3. The curious feature of this reaction is that it involves a nucleus, ^{7}Be, that has a half-life of 54 days, making it an extremely difficult nucleus on which to study nuclear reactions experimentally. However, an experiment has recently been performed with a ^{7}Be beam (O'Malley et al. 2010, p. 15); the result was that any reactions involving ^{7}Be and ^{2}H could not resolve the lithium problem.

Are there other possibilities that might resolve the problem? One suggestion that has been studied by several authors (Bird et al. 2008; Kusakabe et al. 2007; Pospelov & Pradler 2010) involves the existence of a short-lived particle in the early universe that could have become bound to the nuclei that were formed by BBN. The reason that BBN ceases when it does is that the temperature drops to a point at which the colliding nuclides can no longer overcome the Coulomb barriers that they must surmount in order to react. (The Coulomb barrier is the electrostatic voltage barrier that exists between two positively charged particles.) However, assuming the short-lived particle was negatively charged, it would reduce the Coulomb barrier as soon as the universe cooled to the point at which it could be captured into nuclei that existed at that time and thus would permit a short resurgence of nucleosynthesis. In doing so, it was found that the ^{7}Li abundance problem could be solved. Of course, if such a particle does exist, it should be produced in the high-energy particle accelerators that exist around the world, but it has not been seen yet.

To summarize, the current status of BBN is that the predicted abundances of ^{2}H and ^{4}He are in good agreement with those observed in stars or other environments that astronomers have identified as representing early universe abundances. The agreement with ^{7}Li is poor; the predicted value is about a factor of 3 higher than what is observed.

However, people are not made out of hydrogen and helium; we need other atoms like carbon and oxygen. These are made in stars, which is the subject of the next chapter.

References

Adam, R., Ade, P. A. R., Aghanim, N., et al. 2016, A&A, 594, A1

Bahcall, J. N., Serenelli, A. M. & Basu, A. 2005, ApJ, 621, L85

Bennett, C. L., Hill, R. S., Hinshaw, G., et al. 2003, ApJS, 148, 1

Beringer, J., Arguin, J. -F., Barnett, R. M., et al. 2012, PhRvD, 86, 010001

Bird, C., Koopmans, K. & Pospelov, M. 2008, PhRvD, 78, 083010
Boyd, R. N. 2007, An Introduction to Nuclear Astrophysics (Chicago: Univ. Chicago Press)
Boyd, R. N., Brune, C., Fuller, G. M. & Smith, C. H. 2010, PhRvD, 82, 105005
Burles, S., Nollett, K. M. & Turner, M. S. 1999, arxiv:astro-ph/9903300
Chakraborty, N, Fields, B. D. & Olive, K. A. 2011, PhRvD, 83, 063006
Cyburt, R. H. & Pospelov, M. 2012, IJMPE, 21, 1250004
Hamuy, M., Phillips, M. M., Suntzeff, N. B., et al. 1996, AJ, 112, 2398
Hoyle, F., Burbidge, G. & Narlikar, J. V. 1993, ApJ, 410, 437
Jarosik, N., Bennett, C. L., Dunkley, J., et al. 2011, ApJS, 192, 14
Kusakabe, M., Kajino, T., Boyd, R. N., Yoshida, T. & Mathews, G. J. 2007, PhRvD, 76, 121301
O'Malley, P. D., et al. 2010, in Nuclear Physics with Radioactive Ion Beams, Tennessee Technological University Progress Report ed. R. L. Kozub, (Cookeville, TN: Tennessee Technological University), 17
Perlmutter, S., Aldering, G., Goldhaber, G., et al. 1999, ApJ, 517, 565
Pospelov, M. & Pradler, J. 2010, ARNPS, 60, 539
Riess, A. G., Filippenko, A. V., Challis, P., et al. 1998, AJ, 116, 1009
Riess, A. G., Macri, L. M., Hoffmann, S. L., et al. 2016, APJ, 826, 56
Smoot, G. F., Bennett, C. L., Kogut, A., et al. 1992, ApJ, 396, L1
Wesson, P. S. 1991, ApJ, 367, 399

AAS | IOP Astronomy

Creating the Molecules of Life

Richard N Boyd and Michael A Famiano

Chapter 3

What is the Origin of the Rest of the Elements?

3.1 Introduction to Stellar Nucleosynthesis

The hydrogen and most of the helium in the universe came from the Big Bang, but most of the rest of the entries in the chart of nuclides were made in stars. The chart of nuclides contains all isotopes of all elements; it is much more involved than the periodic table, which contains only the elements, especially because different isotopes of a single element might have been made in very different conditions.

A reminder of some basic facts: different isotopes of an element have the number of protons that defines that element, but a different number of neutrons. For instance, helium (He) has eight known isotopes, all of which have two protons, although only ^{3}He and ^{4}He are stable, that is, they have not been observed to undergo radioactive decay. Carbon (C) has 15 known isotopes, from ^{8}C to ^{22}C, of which only ^{12}C and ^{13}C are stable. Isotopes that are radioactive are denoted as radioisotopes or radionuclides. As noted in Chapter 1, ^{14}C is a radioactive form of carbon with a half-life of 5730 years; this is a particularly useful half-life for determining the ages of things that have lived at one time, and therefore contain carbon.

Stellar evolution involves a number of phases, each of which deals with a different set of nuclear reactions and produces different nuclides. To understand how all nuclides are made, one has to follow all phases of stellar evolution. This does not produce all of the nuclides, however; two or more additional processes are needed to synthesize most nuclides heavier than iron. Figure 3.1 gives a color-coded description of where the different elements come from.

The cores of stars are so hot and dense that they can "burn" the nuclei they contain through nuclear reactions to synthesize different nuclei; this process is called stellar nucleosynthesis. Note that "burn" does not refer to chemical reactions; these are nuclear reactions. They generally occur at much higher temperatures, tens to hundreds of millions Kelvin (in this temperature scale, 10 million Kelvin = 18 million degrees Fahrenheit, and room temperature is about 300 K), and they usually

doi:10.1088/978-0-7503-1993-5ch3

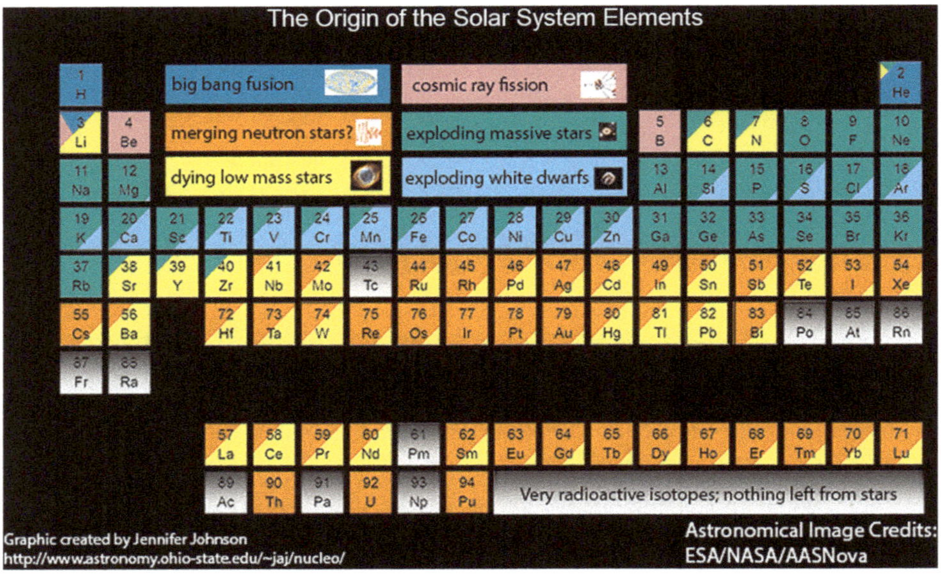

Figure 3.1. Color-coded table of elements from the viewpoint of the production sites. Courtesy of Jennifer Johnson.

produce a lot more energy than chemical reactions. They also proceed much less rapidly at typical stellar temperatures than most chemical reactions we know about from our Earthly laboratory experiments. But that slowness is why stars live for millions or billions of years!

The first stars that formed had to create heavier nuclei—the carbon, nitrogen, and oxygen around which our story revolves—through a myriad of nuclear reactions. What is described below is what goes on in stars that were formed from ingredients that already contained elements that were heavier than hydrogen and helium. These stars were made of the stuff that had been expelled from earlier generations of stars. (Several generations of stars are thought to have produced the material from which our Sun is made.) The first generation burned the primordial stuff, the hydrogen and helium from the Big Bang, but successive generations had some carbon, nitrogen, and oxygen to work with.

3.1.1 Hydrogen Burning

In their first stage of stellar evolution, stars burn hydrogen to convert it into helium. Since most of the helium we see today were created in the Big Bang, the helium abundance has increased only a little as time has gone on due to the helium being produced from hydrogen burning in stars. The second phase of stellar burning, which is discussed below, converts helium into heavier nuclides, but that reduces the helium abundance less than hydrogen burning increases it. Massive stars spend about 90% of their lives in the hydrogen-burning stage of stellar evolution, and the reactions that occur in hydrogen burning depend on the mass of the star. Stars with masses greater than roughly three times the mass of our Sun will have dominant reactions of hydrogen burning that are different from those that operate in less massive stars.

There are two sets of nuclear reactions that describe hydrogen burning: the pp (proton–proton) chains and the CNO (carbon/nitrogen/oxygen) cycles. In our Sun, the pp-chain reactions dominate. Just to give you a flavor of how different these reactions can be, we will write them down. For lower mass stars like our Sun, the reactions go as

$$^{1}\text{H} + {}^{1}\text{H} \rightarrow {}^{2}\text{H} + e^{+} + \nu_e,\ Q = 1.442\ \text{MeV},$$

$$^{2}\text{H} + {}^{1}\text{H} \rightarrow {}^{3}\text{He} + \gamma,\ Q = 5.504\ \text{MeV, and finally,}$$

$$^{3}\text{He} + {}^{3}\text{He} \rightarrow {}^{4}\text{He} + 2\,{}^{1}\text{H},\ Q = 12.839\ \text{MeV}.$$

The nuclei in these reactions were defined in Chapter 2. The symbols e^{+} and ν_e represent a positron, that is, an antielectron (which has the mass of an electron but a positive charge), and an electron neutrino, as defined above. The symbol γ is a gamma-ray. The *Q*-values, that is, the energy released, is given for each reaction. The particles that are important to our story in subsequent chapters will be discussed later on, and we will definitely return to the mysteries of neutrinos. They will turn out to be crucial to our story, but not yet, and not at the energies at which most of them are produced in hydrogen burning. Note that in these equations, the number of baryons (that is, the number of protons and neutrons) is the same on the right and left sides of the equations, so the baryon conservation law mentioned in Chapter 2 is satisfied. The charge conservation law is also obeyed; the amount of charge on the left and right sides of the equations is also equal. For example, in the first equation, the two protons on the left-hand side make the total charge +2. Deuteron, an isotope of hydrogen, has a charge of +1, so the positron is required to make up the additional charge.

In the first equation above, two protons fuse to form a deuteron, a positron, and an electron neutrino, so that conserves the lepton number. In the second equation, a proton and a deuteron fuse to produce a ^{3}He nucleus and a gamma-ray, which is needed to conserve energy, and so forth. The examples discussed so far have defined all of the symbols that we will encounter, and so we will just write subsequent reactions as equations.

In addition to the reactions indicated above, there are a few more that produce nuclei up to ^{7}Be, but those heavier nuclei tend to get returned to lighter nuclei via subsequent nuclear reactions and decays of the radioactive nuclei that are produced along the way. Figure 3.2 shows the three burning chains that convert hydrogen to helium in light stars. The one on the left, the so-called pp-I chain, is the one we described above; the other two are responsible for only a tiny fraction of the Sun's energy output and helium production. It is worth noting that there is a profound difference between the reactions that convert hydrogen into helium in stars and those that operated in BBN. This is a direct result of the neutrons that existed during BBN; neutrons do not exist in stars at a high enough abundance to produce abundant nuclei, except at the time a massive star explodes and becomes a supernova. Thus, the set of reactions of BBN does not apply to stars. Of course, neutrons do exist as stable particles in neutron stars; we will have more to say about those cosmic oddities later on.

In each of the three pp chains shown in Figure 3.2, four protons are fused to produce a ^{4}He nucleus, a particularly tightly bound nucleus. (The importance of this will become clearer in a later section of this chapter.) In the pp-I chain, six protons actually go into that chain, but two of them exist in the final reaction along with the ^{4}He nucleus, so the net number of protons used is four. In the pp-II and pp-III chains (which first make ^{7}Be), the final equations produce two ^{4}He nuclei, but one of them is an input to make ^{7}Be, so the net number of ^{4}He nuclei is one. The pp-III chain actually begins with ^{7}Be, which already has one ^{4}He nucleus and three protons in it, so the one proton that enters that chain produces one net ^{4}He nucleus when ^{8}Be decays. The asterisk indicates that ^{8}Be is produced not in its lowest allowed energy state, but in the next highest energy allowed state.

A discussion of the reactions that occur in stars involve three of the four identified fundamental interactions that mediate stellar evolution. The fourth is what holds

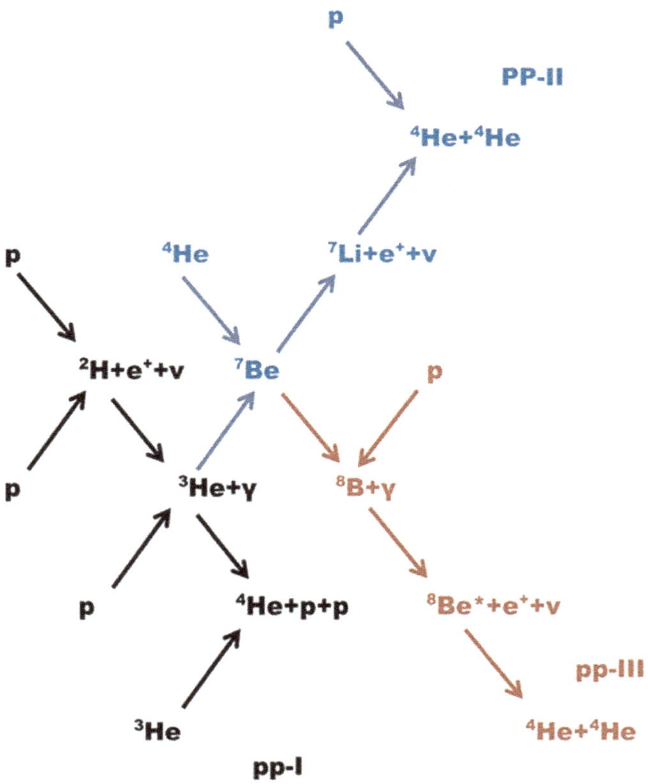

Figure 3.2. The three chains that comprise hydrogen burning in the Sun and in other not-too-massive stars. The pp-I chain produces most of the Sun's energy. The nuclides indicated are actually nuclei. Thus, p is an ordinary hydrogen nucleus, and He is an ordinary helium nucleus..

stars together. They are, in order of their relative strengths: the strong interaction, the electromagnetic interaction, the weak interaction, and the gravitational interaction. For our present purposes, we need only concern ourselves with the first three, and of those, the strong and electromagnetic interactions are much stronger than the weak interaction. (It is aptly named.)

Getting back to hydrogen burning, more massive stars also convert hydrogen into helium, but via a completely different set of reactions than that of the pp chains. These reactions, which comprise the CNO cycles, use ^{12}C as a catalyst nucleus. These nuclear reactions are (the interaction that mediates each one is indicated):

$$^{12}C + {}^{1}H \rightarrow {}^{13}N + \gamma \text{ (electromagnetic interaction)},$$

$$^{13}N \rightarrow {}^{13}C + e^{+} + \nu_e \text{ (weak interaction)},$$

$$^{13}C + {}^{1}H \rightarrow {}^{14}N + \gamma \text{ (electromagnetic interaction)},$$

$$^{14}N + {}^{1}H \rightarrow {}^{15}O + \gamma \text{ (electromagnetic interaction)},$$

$$^{15}O \rightarrow {}^{15}N + e^{+} + \nu_e \text{ (weak interaction)},$$

$$^{15}N + {}^{1}H \rightarrow {}^{12}C + {}^{4}He \text{ (strong interaction)}.$$

In these equations, "C" stands for a carbon nucleus, "N" for a nitrogen nucleus, and "O" for an oxygen nucleus. When a gamma-ray, γ, is produced, the interaction involved is electromagnetic. When an electron neutrino, ν_e, is produced, the interaction is the weak interaction. When there are only nuclei in the equation, the strong interaction mediates the reaction. In this set of reactions, it is seen that the ^{12}C nucleus (which has six protons and six neutrons) catalyzes the capture of four protons (^{1}H), and with two β-decays, produces a ^{4}He nucleus and returns the original ^{12}C nucleus. β-decay is the process by which one nucleus decays to another, emitting a positron, the e^{+}, and an electron neutrino, indicated by ν_e. In β-decay, the total number of baryons—protons and neutrons—in the initial nucleus (for example, ^{13}N—seven protons and six neutrons) is the same as that in the final nucleus (in this case, ^{13}C—six protons and seven neutrons). β-decay proceeds via the weak interaction, which will turn out to be important for our story, but not just because of hydrogen burning. So, it is just another name at this stage. We have also included Figure 3.3 to show these reactions pictorially.

As can be seen from Figure 3.3, the reactions given above are only those of CNO cycles, specifically the loop at the left, but the other cycles do not contribute much energy or nucleosynthesis except at the very high temperatures that occur in massive stars (roughly 10 or more solar masses). Like the pp chains, however, the net result of each of these cycles is the conversion of four protons into a ^{4}He nucleus, although the CNO cycles utilize four proton captures and two β-decays, which differ from the conversion in the pp chains, in which the β-decays are rendered unnecessary by the reaction that produces deuterium.

So why does this CNO cycle burning operate in some stars and the pp chains in others? This depends on the temperature of their environment, and that is determined by the mass of the star; more massive stars are hotter. Since charged particles tend to repel each other if they have the same charge and to attract each other if they have opposite charges, that repulsive force creates the Coulomb barrier, which was mentioned at the end of Chapter 2. Since all nuclei, from protons to beyond lead, are positively charged, they will tend to repel each other. Nuclei are pretty tiny: around 10^{-14} (one one-hundred-trillionth) of a meter. In order for them to fuse, for example, for a proton to fuse with a ^{12}C nucleus to make ^{13}N, the proton and ^{12}C nucleus have to get to within that 10^{-14} m of each other. If the two positively charged nuclei have enough energy, they will be able to get that close. They do not have to have more energy than the height of this barrier. However, quantum mechanical tunneling can get them close enough to fuse even at somewhat lower energies. The temperature of the environment is determined by the motion of the

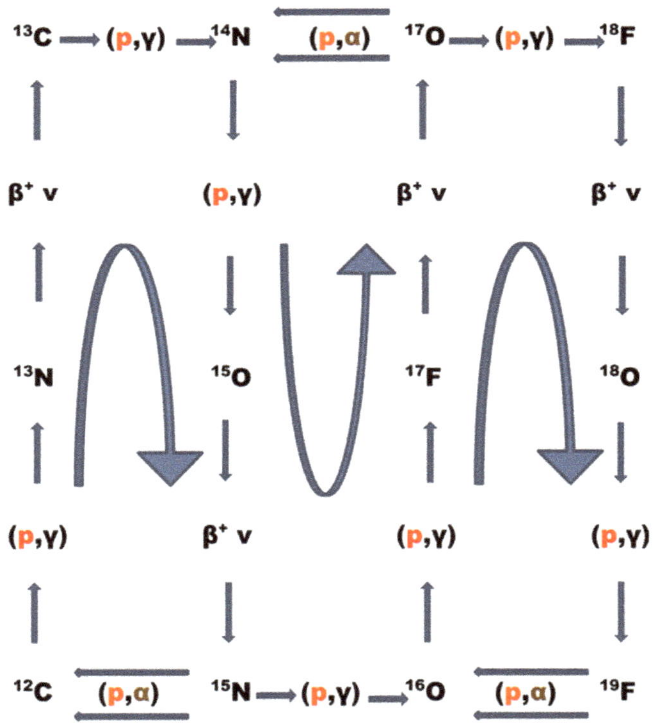

Figure 3.3. The CNO cycles of hydrogen burning. In the cycles on the left and right, the direction of the flow is clockwise, while in the center cycle, it is counterclockwise, as indicated. The cycle on the left, involving carbon, nitrogen, and oxygen isotopes, is the dominant one, although the cycles in the middle and on the right can have important contributions in more massive stars. In these cycles, the β-decays (indicated by the Greek letter β^+), generally occur much more rapidly than the nuclear reactions. Nuclear notation is used here: (p, γ) indicates a reaction in which the nucleus, for example, ^{12}C, captures a proton to make a heavier nucleus (^{13}N in this example), emitting a gamma-ray, indicated by the Greek letter γ. A ^4He nucleus is indicated by the Greek letter α.

particles in that medium, that is, by their energies, and a higher temperature means more energetic particles, so a higher temperature environment will allow its particles to overcome more readily the forces that tend to repel the charged particles. Thus, the reactions that operate in stars will proceed more rapidly at higher temperatures.

The ^{12}C nucleus has a lot more positive charge (six units) than a proton (one unit), so the force of repulsion between the proton and ^{12}C will be a lot larger than that between two protons. Thus, the first reaction in the pp-I chain, $^{1}H + {}^{1}H \rightarrow {}^{2}H + e^{+} + \nu_e$, will occur at a lower temperature than the first reaction in the main CNO cycle, $p + {}^{12}C \rightarrow {}^{13}N + \gamma$. And, since less massive stars have lower temperatures than more massive stars, the pp-chain reactions will dominate hydrogen burning in less massive stars.

But, then, why is pp-chain burning not the primary mode of hydrogen burning in all stars? The answer lies with the $^{1}H + {}^{1}H \rightarrow {}^{2}H + e^{+} + \nu_e$ reaction: it is very slow because it is mediated by the weak interaction. The other reactions operate via the electromagnetic or the strong interaction, both of which are much stronger than the weak interaction, and so will proceed much faster than the $^{1}H + {}^{1}H \rightarrow {}^{2}H + e^{+} + \nu_e$ reaction if the temperature is high enough that they can occur. Thus, if a star is sufficiently massive that the temperature becomes high enough for the CNO cycles to operate, they will take over.

One of the nuclei created in hydrogen burning is crucial to our story, and therefore deserves special attention. This is ^{14}N (seven protons and seven neutrons). Each of the nuclei in Figure 3.3 is made by one process, for example, ^{13}N is made by proton capture on ^{12}C and destroyed by a β-decay to ^{13}C. ^{14}N is the nucleus that is most slowly destroyed in the main hydrogen-burning cycle in massive stars, so a lot of it accumulates as a sufficiently massive star burns its hydrogen. Therefore, a lot of the ^{14}N is likely to be left over after hydrogen burning ends. Much of it will ultimately be expelled into the interstellar medium—the stuff that exists between the stars—following the completion of all stages of evolution of a massive star, which culminates in a supernova explosion. In some stars, i.e., in the Wolf–Rayet stars mentioned in the introduction, ^{14}N can be seen in the winds by which the star expels its outer layer or layers. These stars are especially important to our story, as will be explained later on.

The reason that stars can burn nuclei in nuclear reactions is that their temperature and density are very much higher than the densities and temperatures to which we are accustomed. The temperature at the core of our Sun is 15 million Kelvin, or 27 million degrees Fahrenheit. More massive stars burn their hydrogen at temperatures of 20–40 million Kelvin, or 36 million to 72 million degrees Fahrenheit. The density at the center of the Sun is 150 g cm^{-3}, about 150 times the density of water at Earth's surface, and more massive stars are denser still. So, you would not expect these reactions to occur outside of stars, except in very special situations. It is also not easy to do experiments under conditions that simulate those in stars!

Of course, hydrogen cannot last forever. We believe that our Sun will continue to burn its hydrogen for several more billion years (it has already been at it for 4.6 billion years). More massive stars, however, consume their nuclear fuel more rapidly; a very massive star may live less than a hundred million years (for example, a 40 solar mass star has an expected life of 13 million years). But they will have several more stages of stellar evolution to complete before they end their lives. Some of those stages are very important to our story, so let us spend a little time discussing them. A couple of references on stellar evolution are a textbook on nuclear astrophysics by Boyd (2007) and a review paper by Woosley et al. (2002).

3.1.2 Helium Burning

When the hydrogen in a star's core has mostly been consumed, the energy that was being produced by the proton-induced nuclear reactions decreases. Since the temperature is determined by the motion, or energy, of the particles, if the energy output drops, so will the pressure. As long as the energy produced by the nuclear reactions continues to heat the particles, the resulting pressure will maintain the size of the region of the star in which the nuclear reactions are occurring. But when the fuel that burns during that phase is mostly consumed, the reactions cease, the pressure drops, and the core contracts. As it does, it converts its gravitational potential energy into thermal energy, increasing the temperature. When the temperature in the hydrogen-burning region becomes high enough, the next fuel—helium—will be ignited. There will be plenty of helium in the core; it is, after all, the "ash" of hydrogen burning. The nuclear reactions that dominate helium burning are pretty simple, and they get us to some very important nuclei, so here they are:

$$^{4}\mathrm{He} + {}^{4}\mathrm{He} \leftrightarrow {}^{8}\mathrm{Be},$$

$$^{8}\mathrm{Be} + {}^{4}\mathrm{He} \rightarrow {}^{12}\mathrm{C} + \gamma,$$

$$^{12}\mathrm{C} + {}^{4}\mathrm{He} \rightarrow {}^{16}\mathrm{O} + \gamma.$$

It took nuclear astrophysicists a long time to figure out the first two reactions, since, as noted above, ^{8}Be only lives for 10^{-16} s. Note that this reaction goes both ways, but it will, nonetheless, build up a tiny abundance of ^{8}Be. The density and temperature are sufficiently high in the helium-burning phase that once in a while one of the ^{8}Be nuclei formed for that incredibly short time will capture another ^{4}He nucleus to form ^{12}C. Understanding that reaction required special ingenuity, as the probability for the ^{8}Be to capture the ^{4}He necessitates that a nuclear excited state exist at an energy above the ground state of about 7.6 MeV in ^{12}C—and have very specific properties—for this reaction to proceed. The result is that if the energy of the ^{8}Be + ^{4}He constituents matches this energy, a resonance is created in which the reaction probability is much larger than it would be if the resonant conditions did not exist. Protons and neutrons in nuclei can exist in their special quantum mechanical states of existence, just as we found electrons could in atoms. The ground state is the one with the lowest energy and thus will be the one in which the nucleus will exist unless a particle, for example, a photon, imparts enough energy to promote it to an excited state. The existence of this crucial state in ^{12}C was guessed by Fred Hoyle (mentioned in Chapter 2), simply because stars do make carbon, long before nuclear physicists were able to perform experiments that showed that such a state did exist. It was ultimately found to be at exactly the energy and with exactly the properties that Hoyle hypothesized. Today, this state is commonly referred to as the Hoyle state.

In any event, this stage of stellar evolution makes two very important nuclei: ^{12}C and ^{16}O, both of which are needed to support our Earthly life. In fact, both elements may be essential for life in any form to exist. Thus, both will enter our considerations as major players in later chapters.

3.1.3 Subsequent Burning Stages

If a star has sufficient mass, more than eight times the mass of our Sun, it will go through subsequent stages of stellar evolution that will burn the carbon that is made in helium burning; then the neon that is made in carbon burning; then the oxygen that comprises the ashes of helium burning, carbon burning, and neon burning; and finally, the silicon and magnesium that are made in oxygen burning. As the fuel in each stage is consumed, the star contracts and heats up, so that subsequent stages operate in successively hotter and denser environments.

In each stage, nuclei that are more tightly configured, that is, that have greater binding energy per nucleon, are produced, so that a net positive energy is produced by the next set of nuclear reactions. This can be understood qualitatively by a quick inspection of the graph shown in Figure 3.4, which indicates the binding energy per nucleon for nuclei throughout the periodic table. That is the amount of energy that would be required to completely disassemble a nucleus into its component protons and neutrons divided by the number of protons and neutrons it has. It can be seen that increasing the masses of the nuclei up to the iron–nickel region, around mass 60, will increase the binding energy per nucleon. Since energy is conserved, when less tightly bound nuclei combine to form a heavier nucleus, a net energy will be produced. This is called nuclear fusion.

We will indicate a few of the details of each of these phases of stellar evolution (Boyd 2007; Woosley et al. 2002). Following helium burning, the core of a

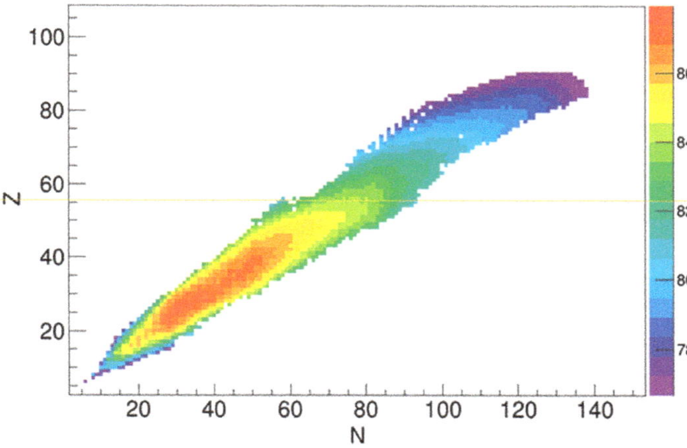

Figure 3.4. Binding energy per nucleon over the chart of nuclides. The values, as indicated by the color-coding indicated on the right, are in kiloelectron volts per nucleon.

sufficiently massive star will contract, become hotter, and will then undergo carbon burning. The reaction that dominates carbon burning is

$$^{12}C + ^{12}C \rightarrow ^{20}Ne + ^{4}He.$$

This reaction conserves the total number of protons and neutrons, 24, but it is not the total fusion of the two ^{12}C nuclei. That would make ^{24}Mg. But that would occur via an electromagnetic interaction, so the reaction that makes ^{20}Ne and ^{4}He, which is governed by the strong interaction, is just more likely to happen.

It appears that we are just moving up the periodic table in choosing our next fuel, in which case ^{16}O would be the next to burn. But it is a very tightly bound nucleus, so ^{20}Ne is the next to burn, and very soon after it gets made, it is destroyed primarily by the reaction

$$^{20}Ne + ^{20}Ne \rightarrow ^{16}O + ^{24}Mg.$$

So now, finally, we get to burn ^{16}O. If it fused, it would make ^{32}S. But once again, that is not the most probable reaction. What actually happens (the strong interaction dominates again) is

$$^{16}O + ^{16}O \rightarrow ^{28}Si + ^{4}He.$$

What follows oxygen burning is called silicon burning, or perhaps a more appropriate name would be silicon melting. The temperature of the stellar core has risen to such a high value, several billion Kelvin, through the succession of contractions and heatings, that there are many highly energetic photons that can interact with ^{28}Si to produce ^{24}Mg and a ^{4}He nucleus. There are also more photons to interact with ^{24}Mg to produce ^{20}Ne and another ^{4}He nucleus, as well as lots of other reactions, too. But those light particles, the ^{4}He nuclei, can get captured on other ^{28}Si nuclei to make ^{32}S, then ^{36}Ar, and so forth. Actually, that is a gross simplification; the reactions that occur are much more complex than that, and they involve other light particles, the protons and neutrons. This has to be described by huge computer codes running on huge computers (Woosley et al. 2002). Regardless, the net result of silicon burning is that it destroys some ^{28}Si nuclei and promotes others until they reach the iron–nickel region.

Looking again at Figure 3.4, one sees that all of the reactions we discussed were able to occur because they fused—sort of, anyway—two lighter nuclei to make a heavier nucleus, which had a greater binding energy per nucleon. Thus, the reaction was exothermic—it produced energy. But once you reach the iron–nickel region, creating any heavier nuclei through this type of reaction will cost you energy, because those reactions are endothermic, that is, the binding energy per nucleon decreases with increasing mass. That is about as far as these phases of nucleosynthesis can go.

3.2 After Stellar Burning

The nuclei around iron and nickel are the primary "ashes" of silicon burning, making that the last stage of fusion burning for a massive star. Thus, the increased energy production that resulted from the contraction following the consumption of each fuel cannot result from burning iron and nickel, and the core of the star collapses, nearly in free fall. The collapse produces temperatures that can increase to several tens of billions Kelvin. At these temperatures, the iron and nickel that were produced in the core will be destroyed by the hot bath of photons that will accompany that temperature. All of the nuclei that the star spent its entire life synthesizing, in the core anyway, will be destroyed in a few seconds.

The core of such a star will collapse until it exceeds the density of an atomic nucleus—2×10^{14} g cm^{-3}—a thimble full of which would weigh 200 million tons! The core of the star will become either a neutron star—a star composed primarily of neutrons (with a small fraction of protons, and maybe some even more exotic stuff) or a black hole. From a personal perspective, a neutron star has a mass somewhat greater than that of our Sun, but a radius of about 10 km (6 miles). This is about the size of the beltway around many cities in the United States, and is also the size of the Yamanote Line, the commuter rail system that encircles Tokyo. A black hole, of course, is the ultimate sink. Once you pass the black hole's event horizon (the point from which nothing can escape the gravitational field of the black hole), you will embark on a very unpleasant journey from which you will never return. At least it will be quick! However, before the collapse to either the neutron star or the black hole can happen, the star will cool its core by emitting a lot of energy. It does so within a few seconds via the neutrinos and antineutrinos mentioned previously, only these are generally at considerably higher energies than those emitted in stellar hydrogen burning and exist in all three flavors, or types. We will return to these neutrinos in a subsequent chapter.

3.2.1 Creating a Core-collapse Supernova

When the core of the star collapses, it does so at such speed that it briefly overshoots the density that a neutron star can sustain. This will produce a core bounce, which will drive an outward-going shock wave that will expel the outer regions of the star into the interstellar medium. At least, that is how astrophysicists describe the supernova explosion mechanism when they could do so in general terms. Although the actual explosion mechanism of such stars is not well understood at present, it is clear from astronomical observations that such stars do explode. Even though the actual explosion details have been difficult for theoretical astrophysicists to figure out, nature has solved the problem.

This explosion will drive the outer layers of the star into the interstellar medium, at least some of the time. The stellar evolution we described above outlines the time evolution of the core of the star, but it also describes the spatial distribution of the different burning stages in a star, which at this stage exist in an onion-skin-like structure. The inner parts are hotter and denser than the outer parts, so when the core is undergoing silicon burning, the next shell out is undergoing oxygen burning, and so forth; thus, the outer layers are burning helium and finally hydrogen. Finally, the burning parts of the star are surrounded by a nonburning envelope. When the star explodes, the expulsion of the outer layers enriches the interstellar medium with the carbon, oxygen, nitrogen, and many other nuclides that were made in the star.

This same star-stuff will form the constituents of future stars and planets, and of people: we really are composed of stardust!

We have included a schematic cutaway picture of a massive star, in Figure 3.5, at that point in its life just prior to its final stage of core collapse and explosion as a supernova to depict what goes on. The onion-skin structure is not something that one should take too seriously; in the olden days when only one-dimensional computations could be done, one simply had to assume that stars were spherically symmetric, as indicated in Figure 3.5. However, now that computers can perform calculations that involve more than one dimension, it is clear that turbulence will produce considerable mixing and distortion of the layers (Arnett & Meakin 2011).

It is worth noting that apparently not all massive stars that end their lives as neutron stars or black holes actually explode in the conventional sense. These silent supernovae apparently emit all the neutrinos that their explosive cousins do, as discussed by Fryer (2009), but somehow fail to emit very much, if any, energy in photons. This may occur because they first form a neutron star as the collapse process begins, and then, perhaps because they began as very massive stars, collapse further to a black hole. The neutrinos will get out; they do not interact much with the matter that they pass through, so they are emitted very quickly. But the photons, the particles of electromagnetic energy, scatter around much more before escaping.

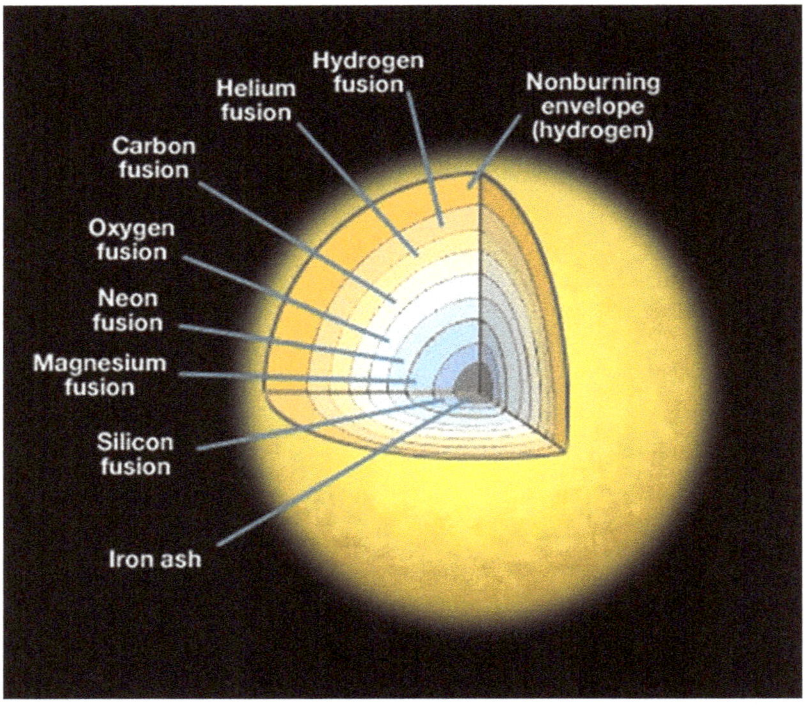

Figure 3.5. The burning stages of a well-evolved massive star, assumed to be initially at least eight solar masses, showing its burning layers, in the usual onion-skin form, just before it undergoes its final stage of core collapse. The figure is not to scale; the nonburning envelope is by far the largest region. Credit: Penn State Astronomy & Astrophysics and Christopher Palma.

There is a good chance they will get swallowed, along with the matter from which they are scattering, by the black hole before they can escape.

3.2.2 Synthesizing the Heavy Elements

Of course, we have not presented a complete picture of nucleosynthesis, because we have not described how elements heavier than iron and nickel are made. Most of those elements are formed by either the s-process, the slow neutron capture process, or the r-process, the rapid neutron capture process. A somewhat less important process is the p-process, as can be seen in Figure 3.6, which shows the abundances of the nuclei made in each process as a function of atomic mass. The synergies among astronomy, nuclear physics, and theoretical astrophysics that were required to produce our understanding of those processes are truly triumphs of science, so we will present some of the details of how they operate in the context of our current understanding. All three processes are still active areas of scientific study. Figure 3.7 shows the paths of the s- and r-processes, the two dominant processes, through part of the chart of nuclides. The paths are defined to be the nuclei that are produced in a process, and they later decay to the stable nuclei seen today. Good references for both s- and r-processes are Cowan & Thielemann (2004) and Seeger et al. (1965), which focus on the r- and s-processes, and Boyd (2007) and Rolfs & Rodney (1988), which also include a description of the p-process.

The s-process occurs mostly during helium burning and synthesizes about half of the nuclei heavier than iron. Its trajectory through the chart of nuclides passes along the neutron-rich edge of the stable nuclei, as can be seen in Figure 3.7. Because the

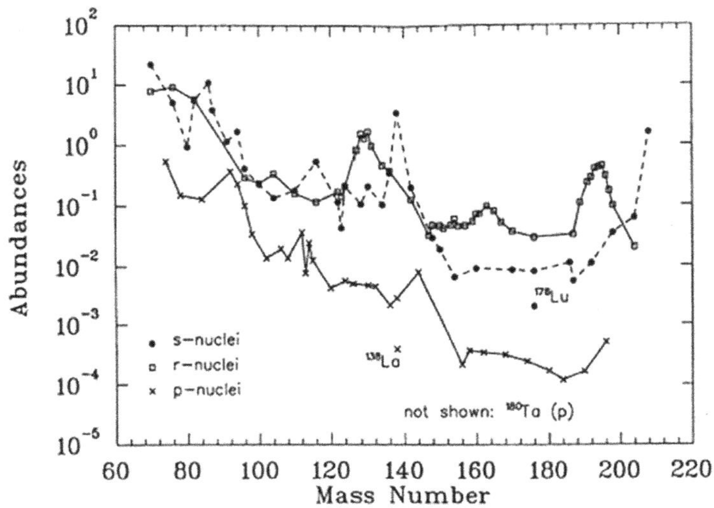

Figure 3.6. The solar system abundances of r-nuclei, s-nuclei, and p-nuclei, relative to Si = 10^6. Only isotopes for which 90% or more of the inferred production comes from a single process are shown. Reprinted from Bradley S. Meyer, "The r-, s-, and p-Processes in Nucleosynthesis," Annu. Rev. Astron. Astrophys. 1994. 32:153-90 Copyright © 1994 by Annual Reviews.

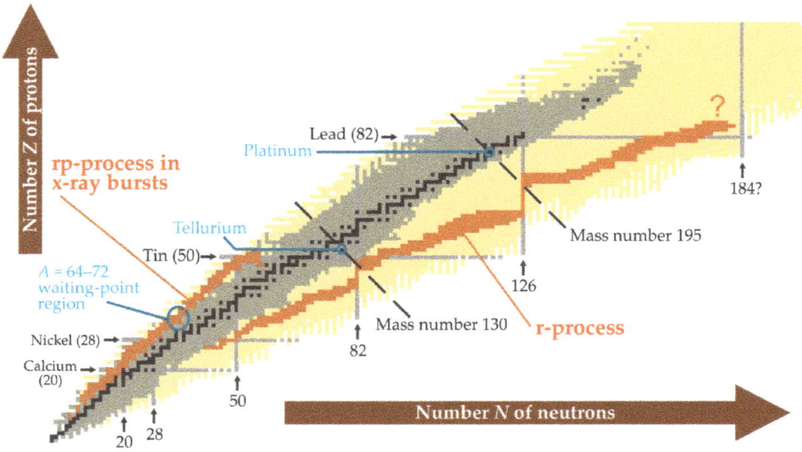

Figure 3.7. Neutron-capture paths for the s- and r-processes through the chart of nuclides. The s-process, which moves along the neutron-rich edge of stability, terminates at ^{209}Bi. Moving past ^{209}Bi requires the r-process. Its path is seen to proceed 10–30 neutrons beyond stability until β-decayed fission and neutron-induced fission terminate it. But after the r-process conditions have ceased, the nuclides that have piled up at the neutron closed shells will β-decay back to the stable nuclides (Thielemann et al. 1983), thus creating the r-process peaks. Another process, the rp-process, is seen to involve lighter nuclides along the proton-rich side of stability. Reproduced from Schatz, H., November 2008, Phys. Today. p. 40, with the permission of the American Institute of Physics. Courtesy of H. Schatz.

neutron captures occur slowly, there is usually plenty of time for any resulting unstable nuclei to undergo β-decay before their next neutron capture, except for a few interesting cases (which often present ways to sample the conditions that prevail during the s-process). Equally important is the fact that, because most of the nuclei involved in the s-process are stable or long-lived, we can study the neutron-capture probabilities on them in the laboratory. These are the dominant unknowns in the s-process, so that means that most of the s-process abundances can be calculated to fairly high accuracy. This will not be the case for the r-process, which we will discuss shortly.

Since these two processes make the heavy nuclides, how do we know what the separate abundances are for them? We discussed some about astronomy earlier in this chapter, and we will have more to say about that in Chapter 4. But for the time being, we will just claim that astronomers can identify elements that are in the periphery of a star from the light that the star emits, or perhaps the light that is absorbed, in the star's surface (Cowan et al. 1997; Roederer et al. 2010; Aoki et al. 2013). As stars evolve and add their newly synthesized elements to the interstellar medium, the abundances of these elements there increases. Then, as new stars are formed from that material, the abundances of those elements will be larger than they were in the preceding generations of stars. Astronomers can tell how far back in time the stuff that they observe in the periphery of a star was made by seeing how much of the key elements—carbon, oxygen, and iron—they observe in that star. Because massive stars make iron, the amount of iron in the galaxy increases with time as more massive stars are created and destroyed. The abundance of iron in a star's photosphere is thus an indication of how many generations of stars have preceded the one in question.

What does this have to do with the abundances of the r-, s-, and p-processes? Since more massive stars go through all stages of their stellar evolution and end their lives more rapidly than less massive stars, the products of nucleosynthesis that are seen in the stars with the lowest abundances of the characteristic elements must have come from the most massive stars. We also know that the most massive stars are responsible for r-process nucleosynthesis (although see below). Thus, the stars with the lowest elemental abundances of key elements, especially iron, are generally thought to have almost totally r-process nuclides. The signatures of the s-process do not seem to appear until later in galactic time, and this is also consistent with our understanding of these processes.

Ultimately, the stars that are formed out of the contents of the interstellar medium have both s- and r-process nuclides. Since, as noted above, one can calculate s-process abundances, one can then subtract those from the total abundances to obtain the r-process abundances. Remarkably, it is predominantly the r-process abundances that astronomers have observed in many of the stars with very low abundances of carbon, oxygen, and iron (Cowan et al. 1997; Roederer et al. 2010; Aoki et al. 2013). This provides a clue that r-process nuclei can be produced early in galactic history when the amount of iron and other medium-mass nuclei was low.

Since neutrons in isolation are not stable particles, they have to be captured pretty quickly after they are produced. That is the case when they are produced in a star; since they have no charge, they do not have to worry about being repelled by the Coulomb barriers of the nuclei with which they interact, and so will fuse rather quickly once produced. Therefore, their capture rate is not the rate that makes the s-process the slow process; it is their rate of production. But that is not all—how the s-process actually occurs is complicated! It has to have stellar regions that are undergoing hydrogen burning in close proximity to regions that are undergoing helium burning in order to produce neutrons. In the hydrogen-burning regions, proton captures produce ^{13}N, which as we discussed above, β-decays to ^{13}C. Now, ^{13}C has to get to a helium-burning region so that the reaction

$$^{13}C + {}^{4}He \rightarrow {}^{16}O + n$$

can produce the necessary neutrons to run the s-process. The nuclear physics is the easy part of this description; it is the hydrodynamics that allows the hydrogen-burning region to produce ^{13}C, and the helium-burning region then to convert that to ^{16}O and a neutron. One also has to describe the details of the dynamical evolution of helium burning; its details change over the time period helium burning operates.

In addition, nature has subjected scientists to yet one more bit of chicanery. In the hydrogen-burning zones that produce ^{13}C, the nuclei therein cannot be processed for too long, or they will create ^{14}N. That is bad for the s-process; ^{14}N is a neutron poison, which will consume neutrons via the reaction $^{14}N + n \rightarrow {}^{14}C + p$. This is governed by the strong interaction without a Coulomb barrier, so it proceeds very rapidly. Thus, the description of the s-process involves huge computers and huge computer codes. And even then the calculations do not always work out. Again, nature knows how to make it happen!

As the s-process moves along its path, it will pause when it gets to the neutron closed shells. Neutrons and protons in nuclei have closed shells just as electrons have in atoms (which is what makes the noble gases—helium, neon, argon, etc.—inert), and it is more difficult to capture another neutron on a nucleus for which the neutron shell has just closed. As the s-process progression pauses, abundance will build up at those nuclei; this produces the s-process abundance peaks at strontium, barium, and lead that are seen in Figure 3.6. Because the s-process is slow, those peaks occur right at neutron closed-shell nuclei.

The heaviest nuclide made by the s-process is ^{209}Bi. If ^{209}Bi captures another neutron, it β-decays to ^{210}Po and then expels a ^{4}He nucleus and ends up back at ^{206}Pb.

However, the heavier nuclides do somehow get made. They are all unstable, but some of them live so long (examples are uranium and plutonium) that we know they did get synthesized. We also know that whatever process synthesizes them must act quickly in order to get past ^{209}Bi, since if one can capture another neutron on ^{210}Bi before it has time to β-decay, one can move on to heavier nuclei. So, this gets us to the r-process. There is one other fascinating aspect of the r-process, which is that it is primary. What that means is that, unlike the s-process, which processes the preexisting seed nuclei and just boosts them to somewhat higher mass, the r-process always seems to start with the same seeds, no matter how many heavy nuclei already existed in the star before it began. Indeed, it appears that the r-process destroys all preexisting nuclei and starts from very basic constituents—protons, neutrons, and ^{4}He nuclei—each time it occurs.

There is not complete agreement as to where the r-process occurs, but the site that has been discussed the most, at least until the merger of two neutron stars was observed in 2017, is core-collapse supernovae. In those situations, there appears to be an intense neutrino wind that blows out from the center of the star, and this produces a high neutron density region. Both seem to be essential for the r-process to work. The maximum temperature that is achieved as the core collapse goes through its successive stages is several tens of billions Kelvin. At that temperature, all of the preexisting nuclei are destroyed. As the star cools, the seed nuclei for the r-process form, making nuclei up to roughly mass 100. The r-process then converts those seed nuclei into all nuclides it forms—all the way to uranium and plutonium and even beyond—within seconds, passing through nuclei scientists have never been able to make, even in our most sophisticated Earthly laboratories. Theoretical astrophysicists have had difficulty showing how core-collapse supernovae really produce the r-process. So, nature seems to have a commanding lead in our understanding of how the r-process actually happens.

As suggested in Figure 3.7, the r-process moves along a trajectory that passes through nuclei that have roughly 20 more neutrons than the most neutron-rich stable nuclei. However, it will also pause at the neutron closed shells that occur along that pathway, and some abundance will build up at those points. This produces the r-process abundance peaks at masses 130 and 195. When the r-process has ended, those neutron-rich nuclei will β-decay back to stability. Thus, the stable nuclei at those peaks are very different from the ones the r-process created; they are much less neutron-rich than the ones along the r-process path. However, those peaks are unquestionably the signatures that the r-process has occurred. The r-process finally ends when the neutron-capture rate drops, either because the neutron abundance drops or because the expansion of the environment results in a density that is too low to support further neutron captures.

The p-process occurs at extremely high temperatures and makes nuclei using the ones that existed before it occurred. The reactions involve almost anything that is exothermic, and this can involve proton captures, neutrino- and gamma-induced reactions, and others. But it also produces considerably lower abundances than the r- and s-processes, as can be seen in Figure 3.6.

3.2.3 Gravitational Waves, Black Holes, and Neutron Stars

An extraordinary scientific event occurred on 2017 August 17. For the first time, the merger of two neutron stars was observed via their gravitational wave signature. This was done by detecting the gravitational waves, the ripples in spacetime, caused by this event. Such mergers may have been detected earlier by gamma-ray observatories, but detecting the gravitational waves was essential to the unequivocal identification of the event.

In order to describe how this was done, we will describe LIGO, the Laser Interferometer Gravitational-wave Observatory. There are actually three similar devices located around the world (Washington and Louisiana in the United States, and Italy). They measure the incredibly tiny strains, or motions, caused in 40 kg test masses (mirrors), suspended by 0.4 mm thick fused-silica fibers, by the interaction of gravitational waves with the interferometer.

A schematic of the LIGO interferometer is shown in Figure 3.8. A laser beam, after considerable amplification and processing, is split into two beams that travel perpendicular to each other through highly evacuated 4 km long arms (encased in concrete to protect them from the elements as well as errant hunters) that have mirrors that reflect the light back and forth in the separate arms 280 times, giving the arms an effective length of 1120 km. The beams are then recombined to see if they have the same phase as when they were separated. Gravitational waves should stretch objects in one direction and compress them along the other, although the dimensional change, or strain, is incredibly small, roughly 1000 times smaller than a proton, or 10^{-17} cm. If no gravitational waves cause any strains of the 40 kg test masses, the two beams will recombine to form effectively the same beam that existed initially. However, if some strain has occurred, the beams will be out of phase with each other. We will talk about phases of waves in a subsequent chapter. For the time being, we will just state that this is what can be observed with the interferometer.

The interferometer's sensitivity is what makes it essential to have more than one such detector. Oscillations can occur from seismic events somewhere on the planet, or even from a large truck passing on a nearby highway. But events like the latter one surely will not appear simultaneously in all three widely separated detectors.

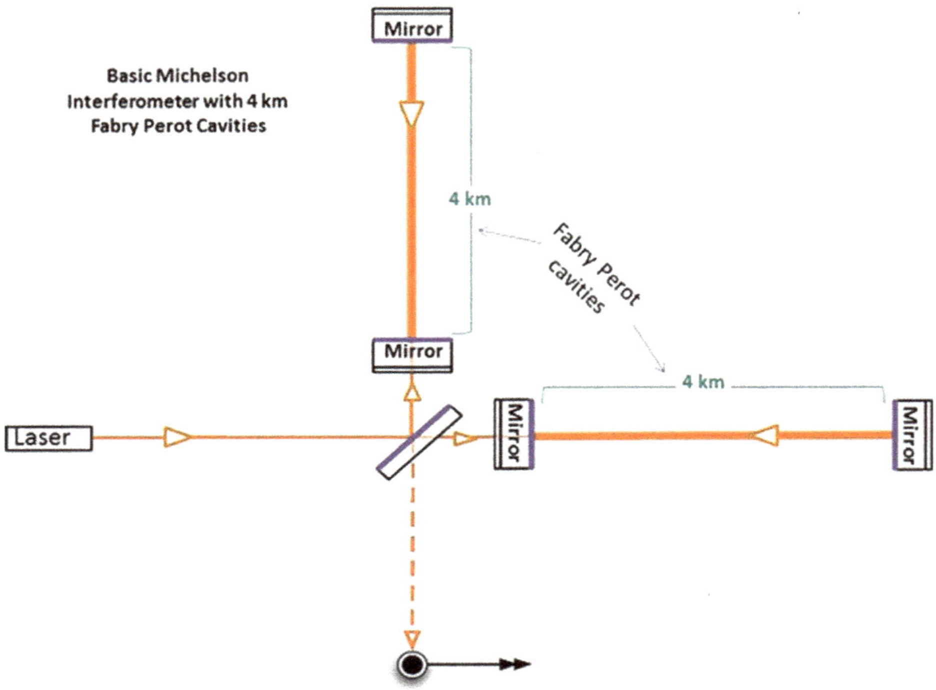

Figure 3.8. The LIGO Interferometer, basically a vastly expanded version of that originally used by Michelson, showing the 4 km long arms located at right angles to each other and the Fabry–Pérot cavities that allow the beam in each arm to bounce back and forth, increasing the effective length of each arm to 1120 km. Courtesy of Caltech/MIT/LIGO Laboratory.

To show how LIGO responds to extreme gravitational events, consider what would happen if two black holes merged. This would begin with them circling each other, as indicated in Figure 3.9. As they radiated away their gravitational energy, their separation distance would decrease, and their rotational speed would increase. This would be expected to produce the sort of oscillatory signal in LIGO indicated in that figure, with the oscillations growing in amplitude and becoming compressed in time as the black holes merged.

On 2015 September 14, the two LIGO detectors that existed at that time observed an event that they interpreted as resulting from the merger of two black holes (Abbott et al. 2016). The LIGO response is shown in Figure 3.10; the similarity between the oscillations seen there and the predicted ones, from general relativity, in Figure 3.9 is striking.

On 2017 August 17, another signal was seen in all three of the world's gravitational wave detectors (Abbott et al. 2017), and it had the characteristics that theorists had thought would be seen from the merger of two neutron stars into a new neutron star or a black hole. As noted above, neutron stars are stars that are incredibly dense. They have masses a bit more than that of our Sun, but are so compressed that they have really huge nuclei that end up about the size of a large city. As their name implies, they are comprised essentially of neutrons.

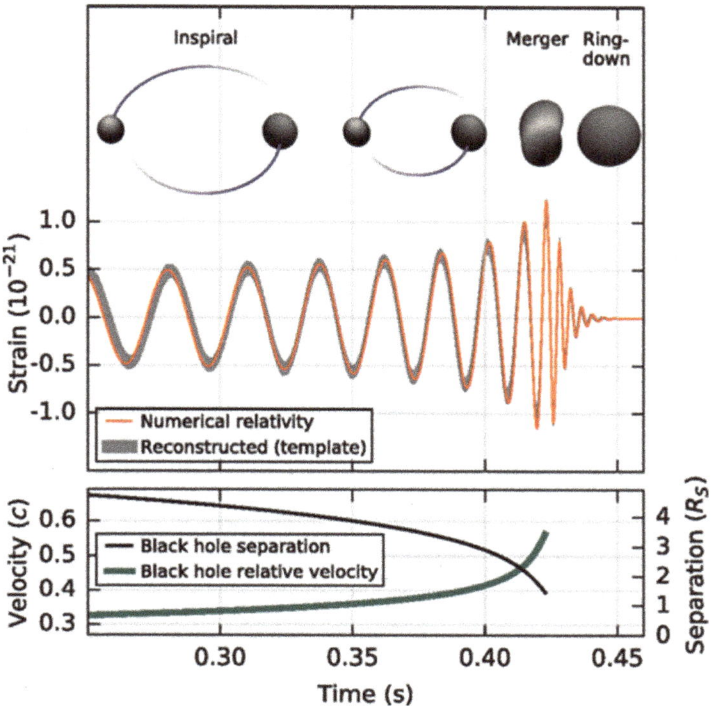

Figure 3.9. Top: estimated gravitational wave strain amplitude from GW 150914, a merger of two black holes, projected onto the Hanford, WA, detector. The inset images show numerical relativity models of the black hole horizons as they coalesce. Bottom: the Keplerian effective black hole separation in units of Schwarzschild radii (RS = $2GM/c^2$) and the effective relative velocity given as a fraction of the speed of light, c. From "Observation of Gravitational Waves from a Binary Black Hole Merger," by Abbott et al. (2016).

The gravitational wave signals were not like anything ever seen before. Previous events, including the black hole merger detailed above, had lasted no more than a few seconds. But the oscillations in the August 17 event lasted for a minute, with oscillations initially at tens of cycles per second and rising to thousands by the end of the event.

However, gravitational waves were not all that was seen. Because the three gravitational wave detectors were spread around the globe, the scientists were able to triangulate their signals to determine approximately where the event occurred—and another completely different type of signal was seen. The *Fermi Gamma-ray Space Telescope* and *INTEGRAL* (*INTErnational Gamma-Ray Astrophysics Laboratory*; and other telescopes) also saw bursts from roughly the same location in space (Goldstein et al. 2017; Savchenko et al. 2017), and they occurred at essentially the same time as the gravity wave detectors saw their signal. The burst was about 1000 times brighter than a typical nova, a different kind of burst, and so was dubbed a kilonova.

This event was also observed somewhat later by the *Chandra* detector (Troja et al. 2017), a spaceborne detector of X-rays. Although the X-rays spanned a fairly large

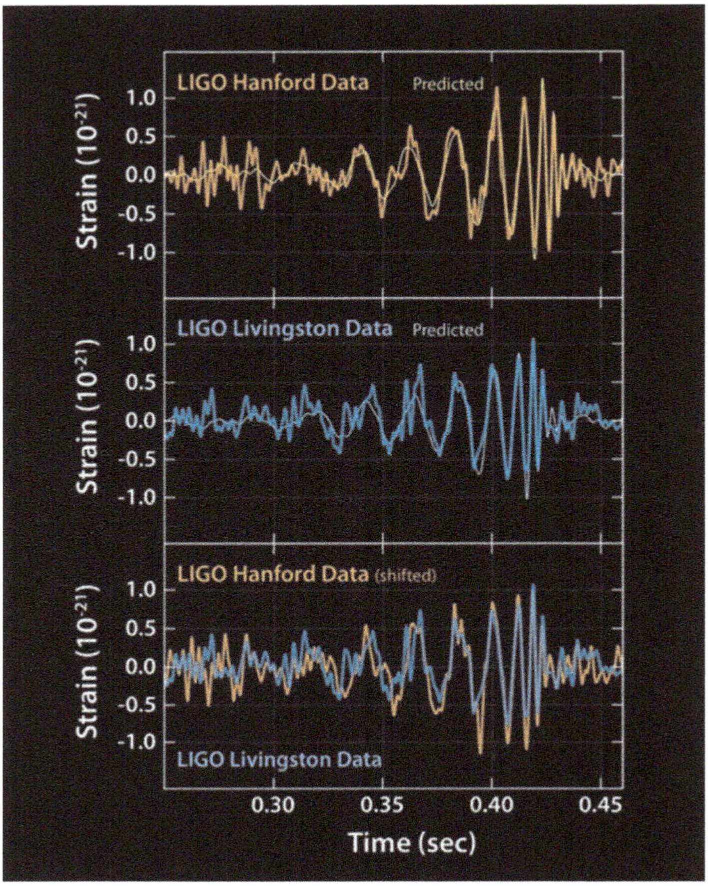

Figure 3.10. These plots show the signals of the gravitational waves detected by the twin LIGO observatories at Livingston, LA, and Hanford, WA. The signals came from two merging black holes, each about 30 times the mass of our Sun, lying 1.3 billion light-years away. The top two plots show data received at Livingston and Hanford, along with the shapes for the waveform predicted for two merging black holes according to Einstein's general theory of relativity. Time is plotted on the abscissa, and strain, the fractional amount by which distances are distorted, on the ordinate. The bottom plot shows the two data sets overlaid, with an appropriate time shift. The LIGO data are seen to closely match each other and Einstein's predictions. Data courtesy Caltech/MIT/LIGO Laboratory.

range of photon energy, it was clear that they indicated that a range of heavy elements from platinum to gold had been formed. This is not unexpected; the combining of huge amounts of extremely neutron-rich matter would be expected to perform an r-process that would be difficult to produce with any other site. There is speculation following the August 17 event that most of the universe's r-process yield comes from neutron star mergers, although this conclusion is not certain. It would help to know how frequently two neutron stars merge, and how well the resulting r-process nuclides are distributed around the Galaxy. Would the products of these mergers be sufficiently frequent and distributed that they would contribute heavy nuclides more or less uniformly? Or would they only affect the local region in which they occurred?

If neutron star mergers do occur frequently, then they would solve the problems theorists have had for decades trying to produce the r-process nuclides from supernovae.

3.3 Evolved Stars

3.3.1 White Dwarfs

What happens to stars with masses less than eight solar masses? These will not have sufficient mass to complete all their stages of burning, and so will not ever become supernovae of the type described above, that is, core-collapse supernovae. They may complete some of those stages and ultimately retire to the status of white dwarfs, stars that are composed either of carbon and oxygen, or of magnesium and neon (depending on how many evolutionary stages they were able to complete), and have masses of not more than around 1.4 times the mass of the Sun. The rest of their mass will be shed in stellar winds, which will enrich the interstellar medium with the elements that were synthesized in the outer shells of the star. A white dwarf will be the final fate of our Sun. These stellar cinders slowly cool in time, but will not perform any additional nucleosynthesis to contribute to the interstellar medium.

Neutron stars are essentially huge nuclei, but white dwarfs are essentially huge atoms. Although they have masses of about that of our Sun, they are about the size of the Earth. Their pressure is maintained by the electrons they contain—through the so-called electron degeneracy pressure—which is the same pressure that maintains the sizes of atoms. This is a result of a principle of quantum mechanics, the Pauli Principle, which only allows one electron to occupy each quantum state. The state of existence of every electron in an atom, or in the white dwarf, is specified, and no other electron will have the same specifications in either object. When you fill up all of the electron quantum states of something that is the mass of the Sun, you get something that is about the size of the Earth. However, there is a maximum mass for a star that can be supported by electron degeneracy pressure; that was shown many years ago by Chandrasekhar. If you want to see how this is determined, see Boyd (2007).

3.3.2 Type Ia Supernovae

However, if the white dwarf happens to be a member of a pair of stars, it may not just sit there for eternity. It may attract matter from its companion, and that may push it beyond the maximum mass that a white dwarf can have. In that case, the electron degeneracy pressure will be overcome, the star will undergo thermonuclear runaway, and the entire star will explode into the interstellar medium. There will be no remnant, no neutron star or black hole. This explosion is also a supernova, but is called a Type Ia supernova; these were used to establish distance scales in the cosmology observations discussed in Chapter 2. Since many white dwarfs are composed of carbon and oxygen, their explosion will also expel those elements that we depend on for life into the interstellar medium. But they will not produce nitrogen;

that is restricted to hydrogen burning, and these stars have very little hydrogen. They also do not produce as many neutrinos as core-collapse supernovae do.

Before leaving the subject of supernovae, we should note that a core-collapse supernova emits of order 10^{53} ergs of neutrinos, or about 10^{57} neutrinos, in a few seconds (an erg is a unit of energy; for reference, a paper airplane in flight would have about 10^5, or 100,000, ergs of energy). Although an erg is a small unit of energy, 10^{53} of them is more energy than our Sun will emit in its entire lifetime. That is all but a small fraction of the energy stored as gravitational energy in the core of the supernova at its formation; the rest, of order 1%, goes into the light emitted and the shock wave that explodes the star. When a supernova occurs, it can outshine all other stars in its galaxy for several months from its photons, and that is just a tiny fraction of its total energy output! We will return to the properties of supernovae in a subsequent chapter.

References

Abbott, B. P., Abbott, R., Abbott, T. D., et al. 2016, PhRvL, 116, 061102
Abbott, B. P, Abbott, R., Abbott, T. D., et al. 2017, PhRvL, 119, 161101
Aoki, W., Boyd, R. N., Kajino, T., Suda, T. & Famiano, M. A. 2013, ApJL, 766, L13
Arnett, W. D. & Meakin, C. 2011, ApJ, 733, 78
Boyd, R. N. 2007, An Introduction to Nuclear Astrophysics (Chicago, IL: Univ. Chicago Press)
Cowan, J. J., McWilliam, A., Sneden, C. & Burris, D. L. 1997, ApJ, 480, 246
Cowan, J. J. & Thielemann, F.-K. 2004, PhT, 57, 47
Fryer, C. 2009, ApJ, 699, 409
Goldstein, A., Veres, P., Burns, E., et al. 2017, ApJL, 848, L14
Meyer, B. S. 1994, ARA&A, 32, 155
Roederer, I. U., Cowan, J. J., Karakas, A. I., et al. 2010, ApJ, 724, 975
Rolfs, C. & Rodney, W. S. 1988, Cauldrons in the Cosmos (Chicago, IL: Univ. Chicago Press)
Savchenko, V., Ferrigno, C., Kuulkers, E., et al. 2017, ApJL, 848, L15
Schatz, H. 2008, Phys. Today 61, 11, 40
Seeger, P. A., Fowler, W. A. & Clayton, D. D. 1965, ApJS, 11, 121
Thielemann, F. K., Metzinger, J. & Klapdor, V. 1983, ZPhyA, 309, 301
Troja, E., Piro, L., van Eerten, H., et al. 2017, Natur, 551, 71
Woosley, S. E., Heger, A. & Weaver, T. A. 2002, RvMP, 74, 1015

Chapter 4

Creation of Molecules in the Interstellar Medium

4.1 The Electromagnetic Spectrum

The preceding chapter described how the elements are synthesized in stars. But how are molecules formed from those atoms in the interstellar medium? There is no doubt that molecules are formed; astronomers have been observing them for many years in the molecular clouds that exist in space via the spectra they emit in the gas phase or from stellar dust grains. Astronomers can analyze the light coming from stars with a spectrograph, an instrument that allows the intensity of light at each wavelength to be measured, and this can often identify the atoms, ions, or molecules that are producing the radiation. Red light has a longer wavelength than yellow light, which has a longer wavelength than blue light, etc. But the colors that we know represent only a tiny component of the electromagnetic spectrum, which ranges from the (very high-energy) gamma-rays, through X-rays, then to ultraviolet light, and on to visible light, then to the infrared, then to radio waves, and finally to the cosmic microwave background radiation discussed in Chapter 2. These are all electromagnetic radiation, and the particles of electromagnetic radiation at all wavelengths, or energies, are called photons.

Very high-energy gamma-rays are produced by high-energy processes in the cosmos or in Earth's atmosphere by some high-energy nuclear reactions. Somewhat lower energy gamma-rays can be produced by some of the same processes that produce the highest energy ones. But they can also be produced by transitions between allowed energy levels in nuclei that have gained energy from a previous interaction to be in an excited nuclear state, either in the cosmos or in Earth's atmosphere. X-rays can be produced by interactions between particles in Earth's atmosphere, or by transitions between allowed atomic states in heavier nuclei with electrons that have been promoted to an excited atomic state by a previous interaction.

doi:10.1088/978-0-7503-1993-5ch4

The protons and neutrons in an atomic nucleus can be arranged in many different ways, and each configuration will have a total energy associated with it that is produced by the forces (strong, weak, and electromagnetic) between the particles. However, not all configurations that one might conjure up are allowed; nuclei are microscopic systems, and as such they are governed by the laws of quantum mechanics, just as the electrons in atoms are. The lowest allowed total energy configuration is called the ground state, while other allowed configurations are excited states. Most nuclei have many excited states, which can decay to an energetically lower lying state, usually with the emission of a gamma-ray to conserve energy.

The same quantum mechanical principles apply to the electrons in an atom, although the configurations of the atomic electrons are governed primarily by the attractive Coulomb force between the (negatively charged) electrons and the (positively charged) nucleus, and by the repulsive Coulomb interaction between the electrons themselves. Atoms also have a lowest energy state, a ground state, and other allowed electronic configurations that will produce a variety of excited states. Decays of the atomic excited states to lower energy states will produce X-rays, ultraviolet light, or even visible photons, unlike the decays of nuclear excited states, which produce gamma-rays.

As can be seen in Figure 4.1, all of the gamma-rays and X-rays, and most of the ultraviolet and infrared photons, will be absorbed as they pass through, or interact with, Earth's atmosphere. This means that if they are to be detected directly, they must be observed by spaceborne detectors. Several such detectors have provided a wealth of data on astrophysical objects. The highest energy gamma-rays, in interacting with Earth's atmosphere as they approach Earth's surface, do produce observable signals from those interactions. Several ground-based observatories have been built to observe gamma-rays by detecting the secondary particles they produce from their interactions with the atoms in the upper atmosphere. These various observatories have allowed astronomers to study some of the most energetic processes that occur in our universe.

As can also be seen in Figure 4.1, Earth's atmosphere becomes transparent to photons in the very narrow window of visible wavelengths. Surely evolution guaranteed the coincidence between this opacity window and the wavelengths that our eyes can detect, which constitute visible light! This is the reason why optical telescopes, which began with Galileo and have now evolved to levels of sophistication far beyond even his imagination, can provide useful data.

However, Earth's atmosphere, while not absorbing those photons, does still have an effect on them. Fluctuations in the atmosphere make the spatial and electromagnetic resolution that one can achieve of much less quality than could be attained without the atmosphere. This is why optical telescopes, such as the *Hubble Space Telescope*, *Herschel Space Telescope*, and others, were put in space.

However, adaptive optics, a technological development that allows fine-tuning of the shape of the telescope mirror on a short timescale to refine its focus, together with the use of either real or artificial guide stars, has pushed the capabilities of ground-based telescopes to nearly the same level of resolution as that of spaceborne telescopes. This is only for visible light, however! Nonetheless, these many

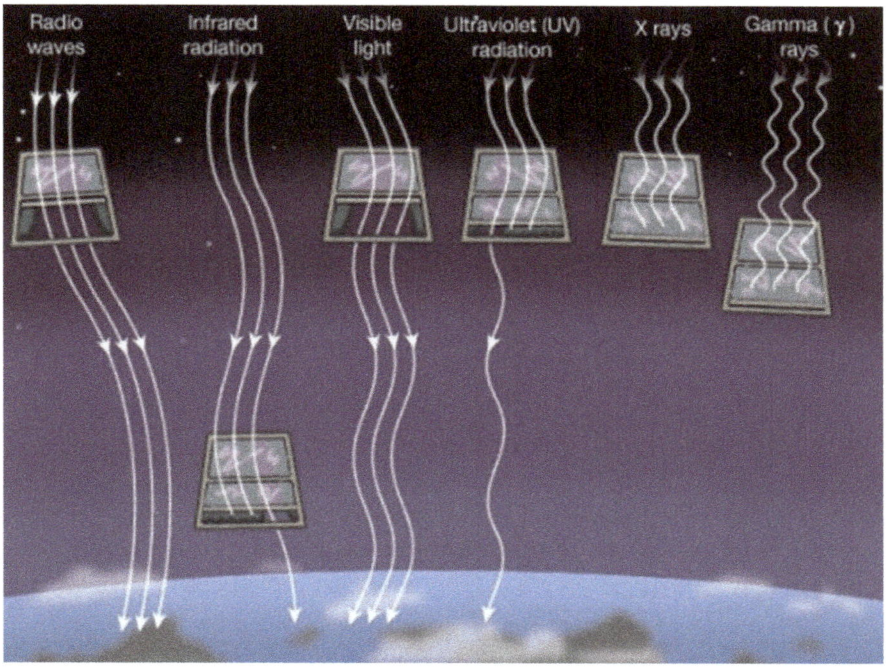

Figure 4.1. Transmission of electromagnetic radiation by Earth's atmosphere as a function of the wavelength of the radiation, with the energy increasing from left to right. Note the rather limited region that allows visible light to be transmitted to Earth's surface. The other region in which transmission occurs is that of radio waves. Various space observatories included in the figure span the electromagnetic spectrum. Credit: Penn State Astronomy & Astrophysics and C. Palma.

developments have allowed astronomers to study the evolution of stars as well as the composition of many objects in the universe.

Between the optical wavelength and the radio wave regions, Earth's atmosphere again becomes nearly opaque; this is the infrared, or thermal, part of the spectrum. However, there are a large number of astrophysical objects that need to be studied at those wavelengths, for example, objects that do produce radiation, but do so at lower temperatures than those that produce optical radiation, so again observatories for photons of those energies have been built and launched into space.

Finally, the emissions from space that are in radio wave frequencies, indicated as radio waves in Figure 4.1, have also provided important astrophysical information. There is a transition between two states in hydrogen atoms that produce a photon with a wavelength of 21 cm; this falls in the radio wave window. Most of the baryonic matter in the universe is actually hydrogen. Thus, this transition, along with radio wave telescopes, has allowed a mapping of most of the matter in the Galaxy, and even well beyond.

Spectrographs provide much higher detail than just basic colors. Each element has a distinctive set of wavelengths that characterize the light it emits or absorbs. Thus, spectral analysis allows the elemental abundances in the periphery of a star to

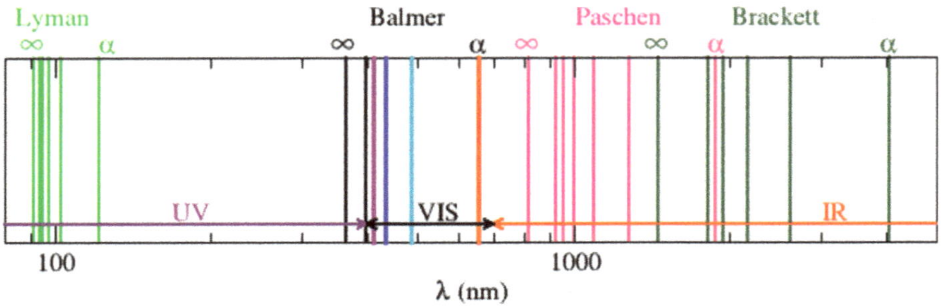

Figure 4.2. Emission spectrum for atomic hydrogen. No other atom would exhibit this same set of emission lines. Redrawn from ChemGuide.co.uk (courtesy of Dr. Clark).

be measured. A spectrum for atomic hydrogen (it often appears in molecular form, that is, in a molecule in which there are two hydrogen atoms) was shown in Chapter 2 to discuss the effects of redshift, but it is shown again in Figure 4.2.

Telescopes detect photons, sometimes from emission from atoms or ions, partially ionized atoms (atoms from which one or more of their electrons have been removed), and sometimes from absorption spectra from the same entities. In this latter situation, the surfaces of stars are backlit by the radiation coming from within the stars. That radiation is fairly uniformly distributed in wavelength and exhibits no characteristic atomic or ionic emission features. The photons from this white spectrum can excite the atoms or ions in the stellar surface, thereby absorbing those photons from the light emitted from that star at the specific wavelength that characterizes the transition.

Molecules also can be identified by their spectra, which, although much more complex than elemental spectra, still provide characteristic signatures. Some that have been positively identified in outer space include molecular hydrogen (H_2), water (H_2O), methane (CH_4), carbon monoxide (CO), carbon dioxide (CO_2), ammonia (NH_3), hydrogen cyanide (HCN), formaldehyde (H_2CO), methylacetylene (CH_3CCH), and many others; the body of observations spanning many decades of effort by astronomers is discussed in the review article by Ehrenfreund et al. (2002). Figure 4.3 shows such a spectrum for hydrogen chloride, HCl, taken in a laboratory experiment. Note how much more complex it is than the spectrum for atomic hydrogen and how much closer together the individual lines are, for the most part. This is because of the many additional possible excitations. Not only are there hydrogen excitations and chlorine excitations, but there are also molecular excitations in which hydrogen and chlorine classically vibrate with respect to each other.

However, just so you can see how different molecular spectra can look when they are viewed in an astronomical setting, we show another spectrum taken with the Submillimeter Telescope, SMT, in Figure 4.4.

4.2 Detecting Photons of Different Energies

Quite different telescopes are required to view photons over the full spectral range shown in Figure 4.1; the wavelengths, and hence energies, span many orders of

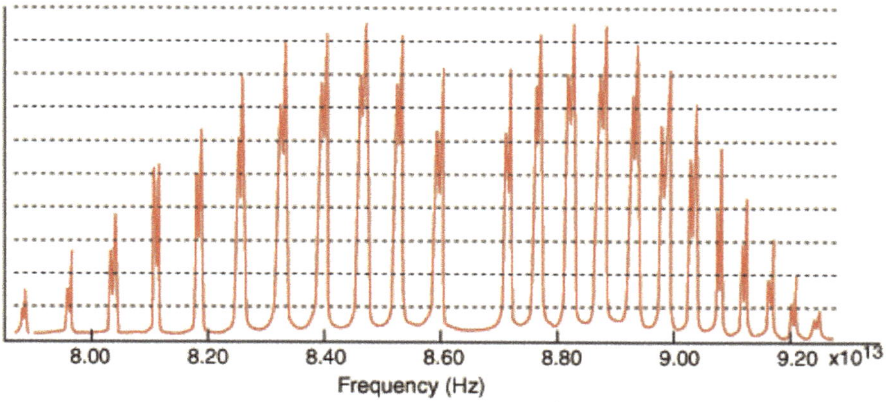

Figure 4.3. Molecular spectrum for hydrogen chloride, taken in a laboratory. The vertical axis is the number of counts per unit time in each frequency bin. Note that the spacings between adjacent lines are much smaller than those for hydrogen, except where the lines in the hydrogen spectrum pile up at high values of the principal quantum number. Courtesy of Rod Nave, HyperPhysics Project, Georgia State University.

magnitude. The different telescopes need to take into account the ways in which photons of that energy interact with matter. The highest energy photons—the gamma-rays—interact directly with the material they strike to produce strong signals in the detectors. However, they cannot be directed by mirrors in the same way that visible light can, since they simply penetrate anything one might try to use to reflect them. The *Fermi Gamma-ray Telescope* is shown in Figure 4.5. The actual detector consists of large blocks of detection material, as indicated in the picture.

X-rays are also difficult to redirect, but they can be reflected through small angles. Thus, the features of X-ray detectors are very different from those designed for gamma-rays. A picture of the *Chandra X-ray Observatory* is shown in Figure 4.6.

The word telescope usually engenders an image of a device with a large dish, now as large as 10 m for optical telescopes (although telescopes with multiple mirrors achieve an effective size larger than that), that serves to reflect and focus visible light onto a solid-state device that converts the photons into electrical signals. In times past, the photons were recorded on photographic plates, but reading the exposures on these could not possibly keep up with the rates at which data are acquired with modern telescopes. Two telescope facilities, the twin Keck telescopes and the Subaru (Japanese) telescope, among others worldwide, utilize adaptive optics. The separate mirrors of the telescope are refocused over time intervals of a fraction of a second using light from either a bright guide star or a synthetic guide star, which is in the form of a laser beam that produces an artificial star from the light subsequently emitted by atoms in Earth's upper atmosphere that were excited by the laser. The Keck telescope is actually two 10 m diameter telescopes, and the Subaru telescope is 8.2 m in diameter. All three are located at an altitude of nearly 14,000 feet (to escape as much of Earth's atmosphere as possible) on Mauna Kea in Hawaii. A figure showing Keck using one of its guide beams is shown in Figure 4.8. Each of the Keck 10 m primary mirrors comprises 36 segments; it is these segments that get adjusted

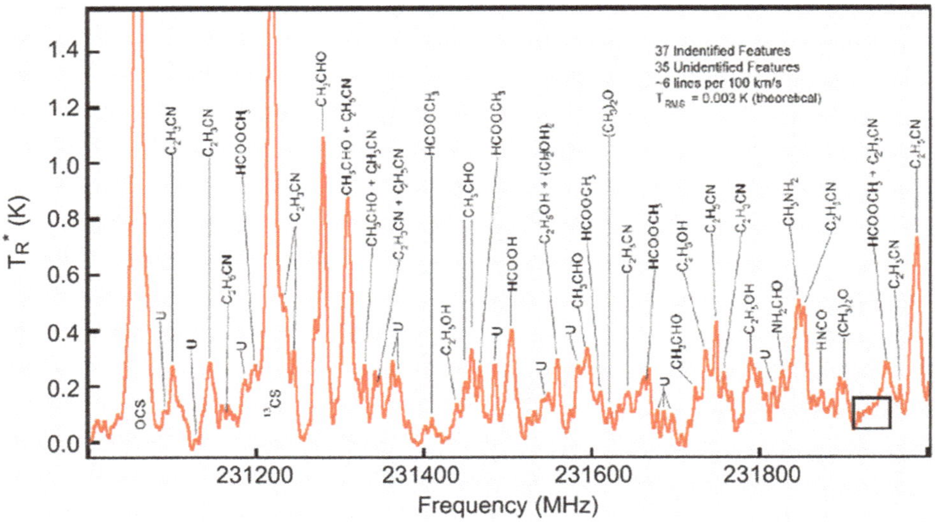

Figure 4.4. Spectrum taken with the Submillimeter Telescope of the Arizona Radio Observatory, showing the spectral complexity with which radio astronomers have to deal. About half of the lines have actually been identified as indicated on the plot. Courtesy of Lucy Ziurys, University of Arizona.

Figure 4.5. The *Fermi Gamma-ray Telescope*. The wings shown in this and other figures are the solar panels that generate power for the telescope's detectors. Courtesy of NASA.

on short timescales to optimize the image of the guide star and, hence, of the objects being studied.

Of course, space-based telescopes do not have to deal with the distortions imposed by Earth's atmosphere, and so the *Hubble Space Telescope*, as well as others, has been designed to operate from space. The downside is that the

Figure 4.6. The *Chandra X-ray Observatory*. The X-rays enter the detector through a thin window in the front. Courtesy of NASA.

restrictions of launch weight limit their size. Nonetheless, they have provided stunning images of cosmic objects that would have been impossible to obtain from ground-based observatories. Figure 4.7 shows the *Hubble Space Telescope*. It is to be replaced by a new orbiting telescope, the *James Webb Space Telescope*, scheduled for launch in 2020.

Infrared radiation requires more specialized detectors but certainly necessitates space-based detection systems because of the opacity of Earth's atmosphere at those wavelengths. A picture of the *Spitzer Space Telescope* is shown in Figure 4.9.

However, other factors need to be considered as well. In order for a telescope to produce a good focus of the photons it receives, it must have gotten careful attention to the precision of its reflective surfaces. For gamma-rays, this is not a consideration, since the gamma-ray will deliver all of its energy to the detector (as long as it is not too close to the edge). However, for optical telescopes, the surfaces must be machined to a small fraction of the wavelengths to be detected. The wavelengths of optical light are several hundred nanometers. Thus, the machining must be good to a small fraction of that: better than 100 nm, or one ten-thousandth of a millimeter. That has to extend over the ~10 m of the mirror's surface if it is a single dish. Finally, the ability of a telescope to localize spatial coordinates of the object that it is viewing is determined by the wavelength divided by the diameter of the telescope, so bigger is definitely better.

Finally, Earth's atmosphere is transparent in the radio wave region, so radio waves can be detected using Earth-based telescopes. For such telescopes, the wavelengths are of the order of centimeters, so the required precision is much less taxing than for optical telescopes. However, radio telescopes need to be huge, hundreds of meters, or even the diameter of Earth (if one combines the signals from many radio telescopes around the globe) in order to localize the object that is being viewed. The radio telescope recently built in China is the world's largest and has a

Figure 4.7. The *Hubble Space Telescope*, shown shortly after it was launched in 1990. This telescope has been one of the primary workhorses for astronomers and has produced a huge number of stunning high-resolution images. Courtesy of NASA.

primary mirror that is 500 m in diameter. The Green Bank Telescope, built in West Virginia, is smaller, but has the advantage that it is steerable. A picture of that telescope is shown in Figure 4.10. The Very Long Baseline Array consists of 10 radio telescopes, plus sometimes several more, ranging in location from Hawaii to Germany. The signals from these telescopes can be combined to produce a telescope with an effective diameter that is nearly the size of Earth. We will return to the possible uses and capabilities of radio telescopes in later chapters.

One especially interesting molecule that has been observed on interstellar grains is methanol (CH_3OH). Methanol and ammonia are generally considered to be the keys to chemical reaction pathways leading to more complex molecules (Ehrenfreund et al. 2002), with ammonia (Shinnaka et al. 1982) being especially critical because it contains a nitrogen atom, which most other molecules do not have. Many heavier molecules, generally referred to as polycyclic aromatic hydrocarbons (PAHs), have also been observed by astronomers. Unfortunately, very few of the PAHs are relevant to our story, although some of them may well be amino acids. But, there

Figure 4.8. The twin Keck telescopes, with one shooting its laser guide star into the heart of the Milky Way, on a beautifully clear night on the summit on Mauna Kea. Courtesy of Andrew Richard Hara/W. M. Keck Observatory.

have not yet been more than one or two convincing astronomical identifications of amino acids, which are the molecules we are primarily interested in. However, the limits on their existence are not very stringent; their non-observation may well just be due to sensitivity limits. One scenario for the creation of those molecules in outer space is that they freeze out in the icy mantles of dust grains, so the more volatile species can be frozen into the grains. The temperatures of these grains in the interstellar medium tend to be around 20 K, 20 degrees above absolute zero.

4.3 Secrets from Meteorites

While the cosmos is apparently teeming with interesting organic molecules, actually finding those that are relevant to life is a bit trickier. However, one laboratory that nature conveniently provides for us—meteorites—makes a completely convincing case that *many of the molecules of life are produced in outer space and that at least some of these can survive the trip to Earth.* This last part is not trivial, as meteoroids passing through Earth's atmosphere (at high speeds!) get heated to temperatures that would destroy many molecules, at least those on their surfaces.

However, there is one type of meteorite that seems especially able to withstand the high temperatures of Earth entry. These are carbonaceous chondrites, which are chunks of matter containing mostly carbon, and which are capable of maintaining

Figure 4.9. Artist's perception of the *Spitzer Space Telescope*. It was put into space in 2003 with a liquid helium supply to cool the detectors enough to allow them to detect the extremely low-energy infrared photons. That has long since evaporated, but some of *Spitzer*'s detectors are still operating and producing useful data. Courtesy of NASA.

Figure 4.10. National Radio Astronomy Observatory's Robert C. Byrd Green Bank Telescope is the world's largest fully steerable radio telescope. Credit: NRAO/AUI/NSF.

themselves and at least some of the molecules they contain even as their surfaces are heated to high temperatures. These are thought to be produced in the detritus of the stellar winds of red giant stars, among other possible places. Red giants are stars that have entered their helium-burning phase.

One large meteorite fell to Earth on 1969 September 28, near Murchison, Victoria, Australia. The Murchison meteorite fragmented as it entered Earth's atmosphere and scattered its pieces over about five square miles. Many of the pieces fragmented again as they hit the ground. However, the chunks that were collected provided scientists with a stunning potpourri of relevant information. The first search for amino acids from the Murchison meteorite (Kvenvolden et al. 1970) produced the result that it contained both left- and right-handed amino acids. But the left-handed ones tended to be in greater abundance, although some amino acids were racemic. A later study (Engel & Nagy 1982) concluded that the nonnatural (on Earth) amino acids tended to have equal abundances of left- and right-handed components (the uncertainties were pretty large), whereas the naturally occurring ones had been contaminated by Earthly amino acids, and so naturally showed a left-handed preference. This result was amplified (Bada et al. 1983) on the basis that contaminants derived from the ground on which the meteorite had landed had skewed the results. It was noted that another meteorite, the Allende meteorite, had demonstrated apparent amino acid contamination to a significant depth inward from its surface, supporting this contention.

Thus, Cronin & Pizzarello (1997) did a subsequent study in which they performed new measurements of the chirality of nonnaturally occurring amino acids, such as isovaline, from the Murchison meteorite. These could not have been contaminated by their Earthly cousins; they do not exist on Earth! What they found was that some of those amino acids had equal amounts of left- and right-handed components, within the uncertainties of the measurements, as had been concluded before. However, others had a definite preference for left-handedness. The Cronin–Pizzarello result was confirmed in a subsequent study by Glavin & Dworkin (2009). One of the many segments of the Murchison Meteorite is shown in Figure 4.11.

An additional study by Martins et al. (2008) produced a stunning result. It showed that several of the compounds crucial to life other than amino acids also existed in the Murchison meteorite. One of the molecules they detected was uracil, one of the nucleobases (see Chapter 1), which are the constituents of human genetic material, ribonucleic acid (RNA). They concluded that molecules that are components of the genetic code were already present in the early solar system and may have played a key role in the origin of life. Although similar chemicals had been previously found in the Murchison meteorite, the possibility of contamination was always present and was difficult to eliminate convincingly. In the study of Martins et al. (2008), isotopic ratios of ^{13}C to ^{12}C were measured in the molecules of interest; these ratios were found to be similar to those found in other extraterrestrial samples, but dissimilar to those of terrestrial origin. So, the result of Martins et al. (2008) has to be taken seriously, although not without question.

Another study of the meteorites was done by Callahan et al. (2011); they also found evidence for nucleobases in meteoritic samples. This study was performed on

Figure 4.11. A segment of the Murchison meteorite. Courtesy of Museums Victoria, https://collections.museumvictoria.com.au/specimens/250.

a dozen meteorites from a variety of locations, ranging from Antarctic ice to the Australian location of the Murchison meteorite. Adenine, one of the nucleobases, was found in 11 of the 12 samples. But more importantly, molecules similar to nucleobases—the so-called nucleobase analogs—that are made along with the nucleobases by the chemical processes thought to occur in meteorites were also found. Some of these are found only rarely on Earth, which suggests that both these molecules and the nucleobases found with them were extraterrestrial. Samples of Antarctic ice and Australian dirt from the location where the Murchison meteorite hit were also studied. They were found to contain far fewer of the nucleobases and nucleobase analogs than were found in the meteorites.

We will expand on the manner in which the nucleobases form DNA and RNA in a subsequent chapter and will comment on the significance of finding the nucleobases in representatives of the cosmos in the final chapter.

There are also results from several other meteorites. The Murray meteorite is of the same vintage as the Murchison meteorite; it was also found (Lawless et al. 1971) to have amino acids, and many of them favored left-handed chirality. The Allende meteorite (www.meteoritemarket.com/AMinfo.htm) fell to Earth in 1969 in Mexico and was also found to contain amino acids. In 2000, a meteorite landed near Tagish Lake in British Columbia, Canada. Analysis of the fragments from this meteorite found a dozen different amino acids. Subsequent analysis, reported by Herd et al. (2011), found that different fragments produced different levels of chiral preference, even for the same amino acids, albeit with large uncertainties. The enantiomeric excess for isovaline was found in one sample to be left-handed at a level of a few percent. Perhaps most notable about this meteorite was that the places where amino acid enantiomeric excesses were large showed notable evidence of aqueous alteration. This conclusion will turn out to have major consequences for our amino acid calculations, presented in Chapter 7, which show much larger enantiomeric excesses if an aqueous environment is present.

Another meteorite that has produced some information about amino acids is the Orgueil meteorite (it was one of the samples studied by Callahan et al. 2011 in their search for nucleobases). Several amino acids were found in that meteorite, and one, isovaline, was found by Glavin & Dworkin (2009) to exhibit significant left-handedness.

4.4 Mining the Comets and Asteroids

Meteorites are certainly an important provider of amino acids to Earth, but another potentially important source of extraterrestrial amino acids could be comets. The nuclei of comets have been described as somewhere between icy dirtballs and dirty iceballs (Ehrenfreund et al. 2002). Their ices contain organic compounds and silicates (inorganic compounds involving silicon). There have been many measurements of cometary constituents, including spectroscopic observations and satellite fly-throughs of cometary tails.

The molecules observed in comets Hyakutake and Hale–Bopp include much water, carbon monoxide, carbon dioxide, methane, ammonia, and many more complex molecules (Ehrenfreund et al. 2002). A search has been made for glycine, the simplest (and achiral) amino acid, but it has not yet been seen. However, as with astronomical observations, the upper limit on its existence is not yet very stringent.

The NASA mission *Stardust* was flown to comet 81P/Wild 2, and at least one amino acid was found in the material that *Stardust* brought back to Earth. The sample that was returned from this comet was sufficiently large that it was possible to measure an abundance for glycine and another amino acid that was determined to be an impurity (Elsila et al. 2009). The glycine, however, was demonstrated convincingly to be from the comet. Unfortunately, it was not possible to measure any amino acid chirality, glycine being achiral.

However, there is clearly a large amount of interest in obtaining new information about cometary constituents; meteorites have greatly whetted the appetites of astrobiologists for new data. One such advance had been anticipated from the return of the mission *Hayabusa* (Japanese for peregrine falcon; Fujiwara et al. 2006; Saito et al. 2006; Yano et al. 2006). It landed on asteroid Itokawa and returned to Earth with samples from it in 2010. It had been hoped that *Hayabusa* would be able to return with some chunks of the asteroid, but the systems of *Hayabusa* that were required for this did not work out. However, it did return with some dust samples from Itokawa, although less material than had been planned for, but apparently still with some cometary stuff. However, it was less than the necessary amount to search for amino acid chirality.

A new mission, *Hayabusa2* (https://en.wikipedia.org/wiki/Hayabusa_2; Zukerman 2010), is currently underway, having been launched in 2014 December. It arrived at its target asteroid 162173 Ryugu on 2018 September 21, and deployed two rovers. This mission is led by the Japanese, but it also involves international collaborators, since some of its instrumentation was built in Germany and France. It has a lander, MASCOT (Mobile Asteroid Surface Scout), that is capable of more than one landing. On its first landing, it will drop off an explosive charge, then circle around to the other side of the asteroid so as to be shielded from the explosion. The crater formed by the explosion will then be hit with a copper penetrator, which will dig down into the dust in the crater to obtain samples from below the surface of the asteroid. This material will be brought back to Earth in 2020 December.

Additional detailed information about amino acids was hoped for from the *Rosetta* mission (see www.esa.int/Our_Activities/Space_Science/Rosetta/Europe_s_comet_chaser and https://en.wikipedia.org/wiki/OSIRIS-REx), a project of the European Space Agency. *Rosetta* landed a module on the nucleus of comet Churyumov Gerasimenko in 2014 and remained in orbit around it for two years. *Rosetta*'s lander had the capability to measure the chirality of the molecules it found. However, the lander bounced upon landing, ended up in a dark place on the comet, and lost its power. Although it did return some information, it was unable to provide any on amino acid chirality. These missions are difficult!

Another mission searching for amino acid chirality on cosmic objects, *OSIRIS-REx* (the *Origins, Spectral Interpretation, Resource Identification, Security, Regolith Explorer*; Thiemann & Meierhenrich 2001) was launched on 2016 September 8. It will land on asteroid 101955 Bennu, a carbonaceous asteroid. It is expected to return to Earth with a sample of the asteroid on 2023 September 24. This asteroid is thought to contain material created early in the solar system, and it is hoped that it will contain organic compounds. The project is being headed by the US National Aeronautics and Space Administration, but its science team has members from the United States, Canada, France, Germany, the United Kingdom, and Italy.

OSIRIS-REx will orbit the asteroid at a distance of 5 km for 505 days to map the asteroid surface and provide the data necessary to choose the optimal place from which to gather samples. The module will land briefly, during which nitrogen gas will blow small particles into a sampler located at the head of a robotic arm. Contact pads will also collect smaller grains. After five seconds, the module will depart. If the first try at collection fails, successive landings (up to three) will be attempted. Then, the robotic arm will retract with its precious cargo, and the module will embark on its return trip to Earth. Figure 4.12 shows *OSIRIS-REx*.

It should also be noted that many comets and meteorites are believed to have hit Earth early in its history, and that each might have carried with it biologically important molecules that could have populated Earth. Indeed, as was discussed in Chapter 1, it is well established that a period of intense meteorite bombardment on the Moon, and presumably also on Earth, occurred early in Earth's history and ended about 3.8 billion years ago (see, for example, Strom et al. 2005 and references therein). As discussed in Chapter 2, the universe, and also our Galaxy, existed for many billions of years before Earth was formed, so there would have been plenty of

Figure 4.12. An artist's conception of the *OSIRIS-REx* spacecraft extending its sampling arm as it moves in to make contact with the asteroid 101955 Bennu. Credits: NASA/GSFC.

time for complex molecules to develop and presumably to have the chirality of their amino acids established. Perhaps they were formed even before Earth existed.

It is interesting to note that primitive life forms are thought to have formed soon (on a cosmic timescale) after the end of the intense meteorite bombardment (Strom et al. 2005; Wilde et al. 2001). However, the first life forms were certainly pretty basic creatures, that is, probably just having single cells, and even those cells did not have all of the components that our cells have. Development of such sophisticated structures would require considerably more time. Multicellular organisms came into being about 1.2 billion years ago, and the divergence of plants, animals, and fungi occurred about 960 million years ago. The planet literally burst with creatures when the Cambrian explosion occurred 540 million years ago. This resulted in the creation of an immense number and variety of living life forms, including the many varieties of trilobites that have been found in the fossil record. Some 250 million years ago, there was a major setback; a mass extinction occurred at the end of the Permian era. Shortly thereafter, though, dinosaurs came into being and ruled the planet until they were wiped out 65 million years ago. *Homo erectus* is not thought to have appeared until about 2 million years ago.

Of course, unless the amino acids that were delivered to Earth were all of the same chirality, it would have been difficult for them to evolve into the homochiral environment that currently exists (although it has been suggested that there may be mechanisms to allow the possibility that a few of them that initially had the "wrong" chirality could be converted to the correct one; Pizzarello & Groy 2011). The comets do have the ingredients to make more amino acids; they have all been found to contain volatile nitrogen and carbon compounds, as well as water ice (Ehrenfreund et al. 2002). Also, as noted above, one comet, 81P/Wild 2, was sampled for amino acids by the NASA spacecraft *Stardust*, which returned to Earth in 2006 with at least one amino acid that is believed to have originated with that comet (Westphal et al. 2014). And we know, as discussed in Section 4.3, that at least some meteorites even contain amino acids, a subject to which we will return in the next chapter. So, there are plenty of objects in the cosmos that could have brought the molecules of life to Earth, provided they all had the same dominant chirality or had undergone the processing necessary to convert them to that chirality.

The meteorites tell us unequivocally that amino acids, among other biologically interesting molecules, have been created in the interstellar medium and that they can survive their treacherous journey to Earth. Furthermore, at least some of them have a preferred left-handed chirality. Given all the meteoritic data, it will be astonishing if forthcoming studies of asteroids change that conclusion. This result is a crucial component to understanding the origin of Earthly amino acids and to our story.

4.5 The Next Huge Step: Forming and Maintaining Life from the Basic Constituents

Obviously, it would be desirable to take the next step, if we can, from the existence of the primitive molecules of life in outer space, or even as they might be created on Earth, to living beings. Although this is an active area of research, it is fraught with uncertainty, since the evolutionary beings that exist now triumphed over their predecessors, thereby eliminating the predecessors and creating missing links. Thus, it is not easy to begin with primitive molecules and figure out how we got from there to where we are now.

Scientists have developed a somewhat simpler procedure for the present. This does not get us to the molecular linkage that we would like, but it is a qualitative time-honored approach to determining the chance that life exists at other places in our Galaxy. This goes by the name Drake equation, devised by Frank Drake in 1961. As described by Davies (2010), it is less of an equation than a way to define and characterize our ignorance. This equation was (as described in Davies' book) established to estimate the number of civilizations from which we might detect radio emissions, as that was the focus of searches for extraterrestrial intelligence at that time. The number N of possible radio-emitting civilizations in our Galaxy is given as

$$N = R^* F_p N_e F_l F_i F_c L.$$

The different factors are

R^*: the rate of formation of Sun-like stars in the Galaxy,

F_p: the fraction of those stars that have planets,

N_e: the average number of Earth-like planets in each planetary system,

F_l: the fraction of those planets on which life could emerge,

F_i: the fraction of those planets with life that evolves to intelligent life,

F_c: the fraction of those planets on which technological civilization and the ability to communicate emerge, and

L: the average lifetime of a civilization that is sufficiently sophisticated to communicate.

Some of the estimates of these factors have huge uncertainties, as you will see from the discussion below. In the end, it will turn out to be difficult to come up with a meaningful estimate for the number of possible radio-emitting civilizations in our Galaxy, simply because of those uncertainties. The point of discussing the Drake equation, though, is to point out the various factors that are involved in creating a radio-emitting civilization.

If life develops on a planet, which is part of a star system, it is assumed that it cannot exist forever, that is, it is born, it lives for a while, then dies, possibly by self-extermination. We will assume that it will not be reborn after it becomes extinct (although if life can be carried to planets from outer space, there is no reason why several lineages of life, each perhaps quite different from the previous, could not evolve on a single planet). If our basic assumption about the singularity of life events on each planet is correct, what we need is the stellar formation rate, R^*, not the number of stars in existence at any given time. Furthermore, we need those stars to be somewhat like our Sun; stars that are too small would never ignite to produce the energy to warm their planets, and stars that are too large would incinerate their planets before any interesting life could get started, or the large stars might not live long enough for intelligent life to develop on their planets. Astronomers have given us a good estimate of R^* (Diehl et al. 2006): about seven Sun-like stars are born per year in our Galaxy.

The known fraction of these stars that have planets, N_p, has grown greatly in recent years due to the planet-detecting satellite, the *Kepler Space Telescope*. Davies gives an estimate of 0.5 for this fraction, but it may now be larger. Even if it were as large as its maximum value of 1.0, though, it would only increase N by a factor of two, and this is an inconsequential uncertainty compared to those associated with the other factors.

The value of N_e, the number of those planets per star that could support life, often referred to as Earth-like, is taken by Davies to be 2. In our solar system, it is generally felt that Earth and Mars could qualify, although the number could be larger when extreme forms of life are included. In any event, we will assume N_e to be 2. Of course, this might require a large protector planet to sweep out the space debris that could destroy life with frequent bombardments, a function that Jupiter performed for Earth. So, perhaps 2 is an overestimate when this is taken into account.

The remaining factors have large uncertainties, and in some cases are subject more to speculation than fact, starting with F_l, the fraction of Earth-like planets on which life actually evolves. As Davies notes, estimates range from 0.01 to 1.0, a range of a factor of 100! If the molecules of life originate everywhere uniformly in outer space, and all that is needed is a cosmic stork to deliver them (and given the above discussion about the results of the meteoritic analyses detailed in Lawless et al. 1971 and www.meteoritemarket.com/AMinfo.htm, such storks may be ubiquitous), then the higher estimate may be relevant. However, only a small fraction of the meteorites are amino-acid-bearing ones, and only a small fraction of those actually contain enantiomeric amino acids. Furthermore, it may turn out that large volumes of Galactic space contain no enantiomeric amino acids at all, even if they do have large numbers of complex molecules, which could make the lower estimate of F_l orders of magnitude larger than it actually is.

It is more difficult to estimate F_i, the fraction of planets on which life forms that will also evolve to intelligent life (by Earth-human standards), and those that achieve life that can develop radio communications capability. The values of these quantities are really unknown and are difficult even to guess with any level of confidence. This is where it would be valuable to know the intermediate stages of molecular evolution from what is delivered to Earth and what is required for modern human existence. Although we might be able to make more meaningful guesses as a result of future microbiological theory and experiments, those simply do not exist at present.

The last entity is the length of time that civilizations live. Davies, an admitted optimist, guesses 10,000 years, but others are less optimistic. What terminates a civilization? It could be nuclear war, a natural catastrophe such as a large meteorite encounter, a massive global climate change, or some other form of Armageddon that ends planetary life. Since the proliferation of nuclear weapons seems to pose a huge threat for mankind and some level of global climate change appears to be a reality, 10,000 years seems like a long time to us. We would opt for 1000 years, but this factor of 10 range (which could be even greater than that) just shows how difficult it is to make these estimates. A related question is even if a civilization lived for 10,000 years, would it be sending out radio signals for that duration? Communications several centuries down the road may take a very different form from the radio waves that are used at present. These last two entries in the Drake equation make it challenging to even try to produce a plausible estimate for the number of possible radio-signal-emitting civilizations in the Galaxy.

However, a factor that might increase L arises from the possibility that several civilizations might occur, successively, on a single planet. This would have to take into account the time it takes for intelligent beings to from the basic molecules of life and to develop radio communication capability. It has taken our civilization a long time for this to happen, but perhaps we are just slower in our development than others might be.

In conclusion, even with the large uncertainties on these entities, it seems likely that there are some other planets in our Galaxy on which life of some form exists, and even at least a few that could send radio signals. The actual number is obviously difficult to determine, but it seems unlikely to us, at least, that it is zero. But it might not be large; the model that we have developed, indeed most of the models that have been developed (see Chapter 9), for the creation of the enantiomeric molecules of life suggest that the planets on which life might exist are in abundance, but the molecules to get life started may not be. Furthermore, the additional accidents that were necessary for us to develop to our present state might not happen frequently.

References

Bada, J. L., Cronin, J. R., Ho, M.-S., et al. 1983, Natur, 301, 494

Callahan, M. P., Smith, K. E., Cleaves, H. J. II, et al. 2011, PNAS, 108, 13995

Cronin, J. R. & Pizzarello, S. 1997, NCimA, 275, 951

Davies, P. 2010, The Eerie Silence: Renewing Our Search for Alien Intelligence (New York: Houghton, Miflin, Harcourt)
Diehl, R., Halloin, H., Kretschmer, K., et al. 2006, Natur, 439, 45
Ehrenfreund, P., Irvine, W., Becker, L., et al. 2002, RPPh, 65, 1427
Elsila, J. E., Glavin, D. P. & Dworkin, J. P. 2009, M&PS, 44, 1323
Engel, M. H. & Nagy, B. 1982, Natur, 296, 837
Fujiwara, A., Kawaguchi, J., Yeomans, D. K., et al. 2006, Sci, 312, 1330
Glavin, D. & Dworkin, J. 2009, PNAS, 10,
Herd, C. D. K., Blinova, A., Simkus, D. N., et al. 2011, Sci, 332, 1304
Kvenvolden, K., Lawless, J., Pering, K., et al. 1970, Natur, 228, 923
Lawless, J. G., Kvenvolden, K. A., Peterson, E., Ponnamperuma, C. & Moore, C. 1971, Sci, 173, 626
Martins, Z., Botta, O., Fogel, M. L., et al. 2008, E&PSL, 270, 130
Pizzarello, S. & Groy, L. 2011, GeCoA, 75, 645 (erratum [75, 6724])
Saito, J., Miyamoto, H., Nakamura, K., et al. 2006, Sci, 312, 1341
Shinnaka, Y., Kawakita, H., Kobayashi, H., et al. 1982, ApJ, 260, 141
Strom, R. B., Malhotra, R., Ito, T., Yoshida, F. & Kring, D. A. 2005, Sci, 309, 1847
Thiemann, W. H.-P. & Meierhenrich, U. 2001, OLEB, 31, 199
Westphal, A. J., Stroud, R. M., Bechtel, H. A., et al. 2014, Sci, 345, 786
Wilde, S. A., Valley, J. W., Peck, W. H. & Graham, C. M. 2001, Natur, 409, 175
Yano, H., Kubota, T., Miyamoto, H., et al. 2006, Sci, 312, 1350
Zukerman, W. 2010, Hayabusa 2 will seek the origins of life in space, www.newscientist.com/article/dn19332-hayabusa-2-will-seek-the-origins-of-life-in-space.html

IOP Astronomy

Richard N Boyd and Michael A Famiano

Chapter 5

Amino Acids, Chirality, and Neutrinos

5.1 A Primer on Amino Acids

Amino acids are the building blocks of life. They constitute the proteins found in nearly every living organism. An example of an amino acid molecule is alanine, shown in two forms in Figure 5.1. Biologically relevant amino acids (those used by living organisms) consist of a central carbon atom bonded to a carboxyl group (COOH: a carbon atom doubly bonded to one oxygen atom and singly bonded to a hydroxyl group, OH) and an amine (NH_2: a nitrogen atom bonded to two hydrogens). These groups are depicted in Figure 5.1. In this way, we can think of an amino acid as a carbon atom forming a central "hub" with a carboxyl group and amine group attached to the "spokes" of this molecule. Such amino acids are called α-amino acids, and the central carbon atom is denoted as the α-carbon. The last group is somewhat arbitrary; it is bonded to a third spoke along with a hydrogen atom.

In its zwitterionic form, the amino acid acquires a significant charge redistribution in which the amine group typically bonds to another proton, while the carboxyl group loses a proton. This redistribution of charge (by moving a hydrogen atom to a different location in the molecule, for example) produces an electric dipole moment in the molecule. Zwitterions can have relatively large electric dipole moments. In solutions, depending on the acidity of the solution, amino acids are generally zwitterionic.

The order of these three groups about the central carbon atom determines the chirality of the amino acid. Left-handed amino acids are arranged with the amine and carboxyl group about the central carbon atom in the clockwise direction (as viewed from the point of view of the extra hydrogen atom), opposite to that of the right-handed configuration. As a notational example, the left-handed orientation for alanine is known as L-alanine, and the right-handed orientation is known as D-alanine. (The L- and D- prefixes have evolved from the Latin terms for left, "laevus," and right, "dextera.") An amino acid's chirality is noted as either left-

doi:10.1088/978-0-7503-1993-5ch5

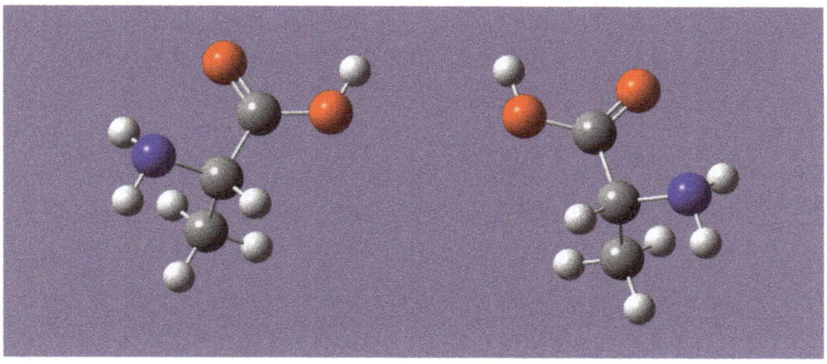

Figure 5.1. The amino acid alanine. The gray, red, blue, and white spheres are the carbon, oxygen, nitrogen, and hydrogen atoms, respectively (taken from the Protein Data Bank; Berman et al. 2000). The two forms shown are L-alanine and R-alanine, the left-handed and right-handed forms of the molecule as discussed in the text.

handed or right-handed, and a chiral state is the handedness of an amino acid. Because these two chiral states are mirror images of each other, they cannot be superimposed onto each other. In a population of amino acids, the relative fractional amount by which one chiral state outnumbers another is given by the enantiomeric excess, *ee*, defined as

$$ee = (N_L - N_D)/(N_L + N_D),$$

where N_L and N_D represent the number of left- and right-handed amino acids, respectively, in an ensemble. Particularly interesting is the fact that amino acids used by life on Earth are exclusively left-handed with only a few notable exceptions among bacteria (Gol'danskii & Kuz'min 1989; Oparin 1957; Haldane 1960; Kvenvolden et al. 1970). Thus, terrestrial life's amino acids (with the exception of the achiral glycine) have enantiomeric excesses of 1.0 or 100%; they are homochiral. Furthermore, the discovery of positive amino acid enantiomeric excesses in meteorites (Kvenvolden et al. 1970; Engel & Nagy 1982; Bada et al. 1983; Cronin & Pizzarello 1997) indicates that amino acid chirality has origins in space and has led to several different theories, some of which we discuss in detail in Chapter 9. As discussed in Chapter 4, analyses of inclusions in meteoritic carbonaceous chondrites have revealed numerous amino acids. Furthermore, some of the amino acids thus synthesized have an excess of a few percent of the left-handed chirality that is observed at the 100% level in Earthly amino acids.

It is generally accepted, from both experiment and theory, that if some mechanism can introduce an imbalance in the populations of the left- and right-handed forms of any amino acid (Gol'danskii & Kuz'min 1989; Kondepudi & Nelson 1985; Mason & Tranter 1985), successive synthesis or evolution of the molecules via autocatalysis can amplify this enantiomeric excess ultimately to produce a single form. What is not well understood is the mechanism by which the initial imbalance is produced and the mechanism it uses to favor the left-handed chirality observed in amino acids. This enigma has been discussed in numerous reviews in the past (Gol'danskii & Kuz'min 1989; Bonner 1991; Boyd et al. 2010,

2011, 2018; Famiano et al. 2014, 2018a, 2018b; Meierhenrich 2008; Rode et al. 2007; Avalos et al. 1998; Guijarro and Yus 2009; Bonner 2000). Solutions include the interactions of molecules with circularly polarized light (Meierhenrich 2008), ultraviolet radiation (Bonner 2000), weak interactions (Vester et al. 1959; Ulbricht & Vester 1962), and other mechanisms. However, many of the proposed solutions to this problem fail in their abilities to produce a sufficient amount of amino acids in a preferred chiral state or to identify sites that would supply the conditions that their explanation requires, or they are unable to produce inclusions of enantiomeric amino acids that could have penetrated Earth's atmosphere for the molecules of life to make it to the surface of Earth.

The question of the origin of biomolecular homochirality has been identified as one of the top questions in science today (Kennedy & Norman 2005), despite the fact that it is nearly 170 years old (Pasteur 1948). Some of the basic features of this problem are summarized below.

5.2 Chirality and Polarization

As indicated in Chapter 1, nature has provided us with an important piece of information that is difficult to reconcile with the spark discharge creation of amino acids: their chirality. We will show why this might be difficult to understand in the context of the Miller–Urey creation mechanism (Miller & Urey 1953, 1959; Johnson et al. 2008).

Chirality refers to the structure of things. As noted in Chapter 1, your left and right hands cannot be translated into each other. But when you look at the mirror image of one hand, now they could be translated into each other, at least in principle. So, your hands are referred to as having mirror symmetry, since they are mirror images of each other. Similar definitions apply to the structure of molecules, most notably the amino acids, which also have chirality. We have shown, in Figure 5.1, the left-handed and right-handed versions of the amino acid alanine. There you can see that amino acids and your hands (and a few other things) share this concept of chirality.

Chirality can also refer to the direction of rotation of circularly polarized light (to be discussed below). The chirality of molecules is closely coupled to the effect the molecules can have on circularly polarized light that impinges on them. For instance, when right circularly polarized light shines on a right-handed amino acid, the light will be affected differently than when right circularly polarized light shines on a left-handed amino acid. Similarly, unpolarized light will behave differently when it is incident on right-handed amino acids than when it is incident on left-handed ones. Such chemicals are called optically active. One way in which amino acids can affect light is through the absorption of polarized light passing through the substance; it will be different for the two polarizations of light. This effect has been used to suggest one of the possible ways in which amino acids developed their specific chirality. In Chapter 9, we will talk about some ways to convert a collection of racemic molecules into one that is enantiomeric. Circularly polarized light is one way to accomplish this. In the meantime, to better understand the relationship between circularly polarized light and molecules, we need to talk more about the basic properties of light.

Light can be characterized by its electric field vector, the representation, in both direction and magnitude, of the effect the field would have on a test charge. Light also has an associated magnetic field vector, but its direction and amplitude are fixed once the electric field vector and direction of propagation of the light are specified. So, we will just characterize light in terms of the electric field vector.

The electric field vector associated with light is an arrow that points in the direction of the electric field while the light particle, the photon, is moving through space. The field will oscillate in the positive and negative directions, as indicated in Figure 5.2. That figure indicates the oscillations of both electric and magnetic fields as the wave propagates through space. However, the field also oscillates in time; the figure could equally well represent what you would see if you sat at a fixed location and watched how the electric field oscillated as the photon passed by you. The wavelength of the oscillations is the distance over which one cycle occurs. The period is the length of time over which one complete cycle occurs. Thus, the horizontal axis could be labeled as either space or time. The arrows could be oriented so they point to the left or to the right as the electric field oscillates, as shown, or they could also be oriented to oscillate up and down. Or they could occur in any other plane. In any of these cases, the light is said to be linearly polarized.

Suppose that we add up–down and left–right electric field vectors of the same strength (or length) so that the one going up and down was at its maximum up direction exactly when the one going left and right was maximum to the right. The up–down and left–right electric field components are then said to be in phase. The resultant electric field direction is at 45° to both the up and down, and left and right components. This light is also linearly polarized in the up–right and down–left directions. However, the direction of the arrow is always in the plane that is 45° to both the vertical and horizontal axes. The amplitude of the total electric field, that is, its maximum value, increased a bit from the vector addition.

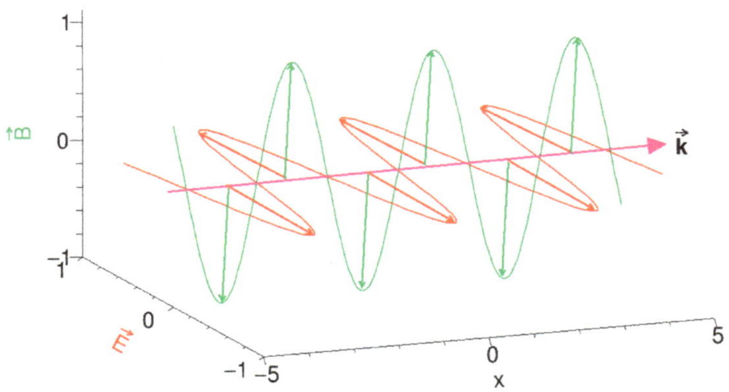

Figure 5.2. Linearly polarized light, with both electric (red) and magnetic (green) fields, which are perpendicular to each other, indicated. The direction of travel of the wave is indicated by the vector ***k***.

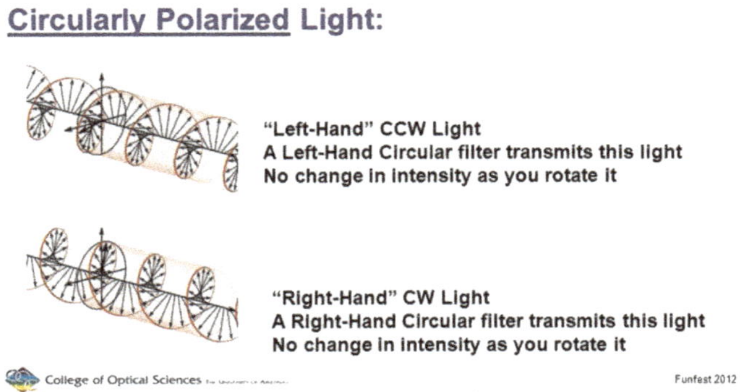

Figure 5.3. Electric field vectors for circularly polarized light. Courtesy of the College of Optical Sciences, University of Arizona.

When we add two electric field vectors so that the up–down vector is maximum when the left–right one is zero, we get what is referred to as the vectors being out of phase. In this case, the electric field vector appears to rotate, as indicated in Figure 5.3. If the maximum values of the up–down and left–right vectors are the same, the length of the total electric field vector will be constant. In that figure, you are looking in the direction of propagation of the light. By changing the relative timing of the up–down and left–right components, we can make the direction of rotation go in the other direction, as shown in Figure 5.3. The one that appears to have the electric field direction rotating clockwise (or counterclockwise) is referred to as right (or left) circularly polarized light. If this condition of the up–down and left–right vectors being 90° out of phase is changed to some other angle, or if the magnitudes of the x and y electric field components are not the same, the footprint of the electric field vector on the x–y plane will form an ellipse; the light is then elliptically polarized.

5.3 Circularly Polarized Light and Molecular Chirality

Circularly polarized light and molecular chirality are correlated. This correlation might take the form of attenuation of one form of circularly polarized light and transmission of the other, or it might be that the electric field direction gets rotated in different ways by the two molecular forms. Suffice it to say that chirally selected molecules do impose different effects on the two forms of circularly polarized light, and that these effects are related to the structure of the molecules.

It is truly remarkable that all amino acids that are naturally occurring on Earth, except the achiral glycine, are left-handed. It is also a problem for the Miller–Urey interpretation of the origin of life, since their experiment produced equal numbers of left- and right-handed amino acids. So how did the amino acids all become left-handed? We will discuss several possible solutions to this problem in Chapter 9. We will also discuss a recently suggested solution that circumvents many of the problems with the other solutions.

5.3.1 Chiral Selection

How might chiral selection of molecules occur? This will require that the different chiral forms of molecules be distinguishable, in energy, or in spacetime, or in some other way. To explore this, we go back to Michael Faraday (1846), who, in 1846, noted that conductors in an external changing magnetic field will tend to act as if they had an induced current. This results from the fact that electrons moving in a magnetic field feel a force that is perpendicular to the field. Thus, they will tend to move in circles, with the plane of the circles perpendicular to the direction of the magnetic field. Although this would not of itself produce a different effect on left- and right-handed molecules, it does create that possibility. This effective rotating electron ring will produce a magnetic moment, which describes the interaction of the molecules with the external magnetic field.

Additionally, particles and nuclei can also have their own intrinsic magnetic moments. In particular, if a particle, or a nucleus, has an intrinsic spin, a degree of freedom that makes the particle or nucleus act as if it has a rotating ring of charge (it does not; it only acts that way), it will have a magnetic moment. The vector that represents the magnetic moment points in the direction perpendicular to the plane in which the charge rotates—or appears to rotate. Nuclei like ^{12}C and ^{16}O have spins of zero if they are in their ground states, their lowest energy states, so they are not candidates for this effect. However, ^{14}N has a nonzero spin, which will interact with an external magnetic field and with the electron orbitals. Thus, ^{14}N is crucial to the model we are describing; the effect it produces could not exist for either ^{12}C or ^{16}O. Also, ^{14}N is an atom that is common to all of the amino acids, so its spin may impact amino acid chirality.

The effect of nuclear spin on a molecule in a magnetic field was described in Buckingham's (2004) work, which was designed to explain the effects of chiral discrimination in nuclear magnetic resonance studies, and was extended by Buckingham & Fischer (2006) and directed in different ways by others (Bonner 2000; Faraday 1846). However, the effects Buckingham described for molecules in the magnetic field of a laboratory magnet would also apply to molecules in outer space if they are subjected to both external magnetic and electric fields.

Nature conveniently produces the conditions that give rise to the Buckingham effect in amino acids. These arise in neutron stars or black holes as they are forming when a core-collapse supernova occurs. The magnetic field close to the surface of a neutron star can be billions of times stronger than the strongest magnetic field that scientists have produced in laboratories on Earth, and so we might expect these fields to produce extraordinary effects. In addition, electric fields can be generated in the reference frames of moving bodies in the vicinity of these magnetic fields through what is known as the motional electromotive force, EMF, which is a result of the Lorentz force. These electric fields force opposite charges in opposite directions. Both this electric field and the magnetic field are important, but they require one other ingredient to produce amino acid chirality, and supernova collapse or neutron stars supply this ingredient in abundance. These are electron antineutrinos, mentioned in Chapter 1, which are emitted in copious numbers from a supernova as a star collapses to a neutron star or a black hole, or possibly from a cooling neutron star.

What Buckingham showed is that, in the presence of external electric and magnetic fields, the local magnetic field at the nucleus of the molecule differs slightly from the external magnetic field. This difference depends on the orientation of the magnetic field with respect to the electric field and on the orientation of the magnetic field with respect to the molecule. Because molecules have electric dipole moments (produced by their charge distributions), the entire molecule can be oriented in the electric field. Electric dipole moments depend on the electronic configuration, which means that they depend on the molecular chirality (see Figure 5.1). Thus, the shift of the magnetic field at the nucleus also depends on the molecular chirality. A strong external magnetic field, together with the nonzero-spin nucleus of ^{14}N, will allow nature to distinguish between left- and right-handed molecules. But how would this relate to the chirality of the molecules? This is discussed in detail in Chapters 6 and 7, and in the appendices. First, we need to review some basic quantum mechanics.

5.3.2 Some Basic Quantum Mechanics

Atoms and molecules must be described from a quantum mechanical point of view, so some background on quantum mechanics as it applies to the components of atoms will be provided. If you are quantum mechanically conversant, you may want to skip a few paragraphs.

An atom will have a magnetic moment that will depend on its total angular momentum of $J\hbar$, where J can be either an integer (0, 1, 2, etc.) or a half-integer (1/2, 3/2, 5/2), depending on whether the total angular momentum of the atom is an integer or half-integer, and $\hbar$ is Planck's constant divided by 2π. The interaction of the magnetic moment of the atom with an external magnetic field will produce a separation in energy of its magnetic substates, which can be observed by detecting the photons produced or absorbed in the transitions between them.

The magnetic substates are characterized by a quantum number that ranges from the value of the total angular momentum to minus the total angular momentum in units of one. For example, if $J = 2$, the magnetic substate quantum numbers, m_j, will be −2, −1, 0, 1, and 2. If $J = 3/2$, the magnetic substate quantum numbers will be −3/2, −1/2, 1/2, and 3/2. These represent the different orientations of the vector J with respect to the direction of the magnetic field, usually taken to be the z-axis. However, in quantum mechanics, the magnitude of the total angular momentum vector is greater than the magnitude of any of its substates.

For the $J = 3/2$ case, the total angular momentum is a bit larger than 3/2, the length of its maximum projection. An external magnetic field breaks the spherical symmetry of the system. Since the total angular momentum is greater than the length of any of its components, it can never be perfectly aligned with one of those components, a direct result of the quantum mechanical nature of spin and angular momentum. A visualization of this concept is given in Figure 5.4. The length of the total angular momentum never changes (angular momentum is a conserved

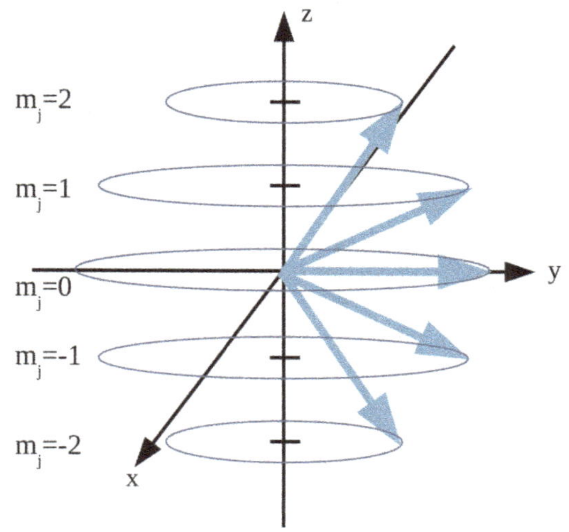

Figure 5.4. Angular momentum vectors in quantum mechanics for the $J = 2$ angular momentum case. The values of the quantum numbers m_j that characterize the magnetic substates are a projection of the total angular momentum onto the z-axis. The angular momentum vectors are the light blue lines. The x- and y-components of the angular momentum are arbitrary in this case.

quantity), but its z-component is always quantized and can change as transitions occur between substates.

In a magnetic field, the angular momentum of a particle can be coupled to its spin, S, through the following relationship:

$$\mu_S = gqS/2m = \gamma S,$$

where q is the charge of the particle, m is its mass, and g is its gyromagnetic ratio, which is intrinsic to the particular particle or atomic nucleus in question. While electrons may have the same angular momenta as nuclei, their magnetic moments are much larger. As a result, the energy states of an electron are different from those of nuclei in a molecule. Further, much like atoms, many molecules that we study are closed shell, with the electrons pairing up to fill the states that make up the molecular bonds. This means, among other things, that the total electronic magnetic moment is zero because the sum of the electronic spins and orbital angular momenta add to zero. In this case, we will concern ourselves with magnetic effects on just the nuclei of molecules, revisiting some special cases later on.

The nuclei of amino acids will necessarily have many total angular momenta, since each amino acid is composed of many different atoms. For each amino acid, its total angular momentum will be the vector sum of the angular momenta of each of its constituent electrons and nuclei. ^{14}N is not the only atom with a nonzero angular momentum. Hydrogen nuclei, for example, also have a nonzero angular momentum from the spin of its proton (1/2). However, the effects of the hydrogen atoms in the molecule on molecular chirality selection are expected to be much smaller than those associated with ^{14}N. This is due to an unusual selection effect associated with the alignment of the spins of ^{14}N, which we will discuss subsequently, with those of the neutrinos that process the molecules. This effect does not exist for H nuclei.

The thermal equilibrium that would exist in any environment will dictate that the numbers of molecules in each of the magnetic substates (that is, their populations) in any group of such atoms would be about the same for most temperatures, assuming that the molecules have time to come to thermal equilibrium.

However, when an external magnetic field is applied, nuclei tend to become oriented. The Buckingham effect would create a chirality-dependent redistribution of the electron current density in the molecule, which would produce magnetic field shifts at the nuclei driven in opposite directions for the two chiral states. The situation is illustrated in Figure 5.5. On the left, the populations of the magnetic substates are seen to be equal prior to the condition that generates a strong magnetic field, for example, a supernova going off, and the magnetic field is very small. However, in the presence of the subsequent strong electric and magnetic fields, the left-handed molecules have their distributions driven in one direction (positive or negative) from their thermal equilibrium positions without the electric field, and the right-handed ones into the other direction, as indicated in the figure. The effect is greatly exaggerated in the figure.

How this distinction might manifest itself in a way that would produce an enantiomeric excess will be discussed in Chapters 6 and 7.

5.4 Basic Features of the SNAAP Model

The Supernova Neutrino Amino Acid Processing (SNAAP) model (Boyd et al. 2010, 2011, 2018; Famiano et al. 2014, 2018a, 2018b) requires properties that might result from a supernova, so let us begin by describing what those properties are and how they affect chirality selection. When a core-collapse supernova occurs, the core of the star contracts into either a neutron star or a black hole. However, in the latter scenario, the

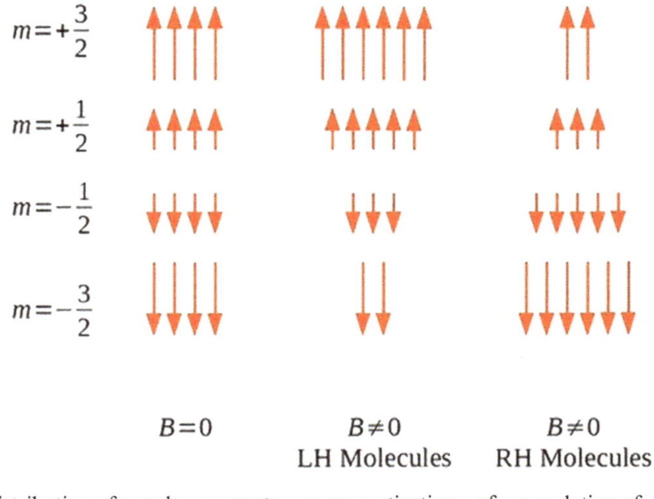

Figure 5.5. The distribution of angular momenta—or magnetization—of a population of atomic nuclei in the presence of no magnetic field (left) and with a magnetic field (right).

supernova may first contract to a neutron star, then collapse to a black hole after some material that was initially blown outward falls back onto the neutron star, causing it to exceed the maximum mass that a neutron star can have. That mass is determined by the degeneracy pressure of the neutrons and the few protons contained in the star. The collapse to a black hole might take a short time, on the order of seconds, but that is plenty of time for a lot of neutrinos and antineutrinos to be emitted. The general result is that, independent of the final state of the star, a lot of neutrinos will be produced as the final contraction occurs, and the nascent neutron star or black hole will probably establish an enormous magnetic field. Both are essential to the SNAAP model.

What happens if the initial star is sufficiently massive that it first collapses to a neutron star and then collapses again to a black hole? As noted above, this does not take very long, but the collapse affects things other than the neutrinos in different ways. Some parts of the star that exploded outward may be able to remain outside the black hole. But other parts might be infalling when the core object became a neutron star and would not be likely to escape. What happens to all of the photons that were spewed out from the center of the star when it exploded? The photons scatter from the matter that surrounds the core, and it takes them perhaps an hour to appear to the outside world. But that is a very long time compared to the collapse time of the core, so those photons are likely to be trapped by the infalling matter and to disappear inside the event horizon of the black hole. These supernovae will not produce much of a show compared to those that do not collapse to black holes; indeed, they would not be seen at all via their photons. These are called silent supernovae because they produce little, if any, light emission.

Supernovae that do explode are such gorgeous objects that we thought it appropriate to include a picture of one. Figure 5.6 shows Cassiopeia A, which was observed to have exploded 11,000 years ago. The picture shown is actually a composite of the radiation detected with two orbiting observatories, *Spitzer* and *Chandra*, and the radiation in the optical wavelengths detected with Earth-based telescopes.

However, there is another aspect of this collapse that is important to our story. This is the incredibly strong magnetic field, mentioned above, that is often generated in the collapse to a neutron star or black hole. The magnetic field lines that are produced around a neutron star are shown in Figure 5.7. It is a dipole field, so the field lines emerge from the north pole of the star and return to the south pole.

Since the magnetic field lines are outgoing at one side and incoming at the other side, they form closed loops. The magnetic field from the neutron star will interact with the magnetic moment of ^{14}N. Since the magnetic moment is related to the spin, the directions of the spins of ^{14}N will be along the magnetic field lines, an essential feature of the SNAAP model. The directions of the neutrino spins, also indicated in Figure 5.7, will be outgoing; they are unrelated to the direction of the magnetic field.

In the SNAAP model, it is assumed that amino acids are produced and constrained in meteoroids, since this has been found to be the case for those that made their way to Earth. When a supernova explodes, whatever meteoroids are reasonably close to the nascent neutron star will be subjected to the strong magnetic field that results from the core collapse and to an intense flux of electron

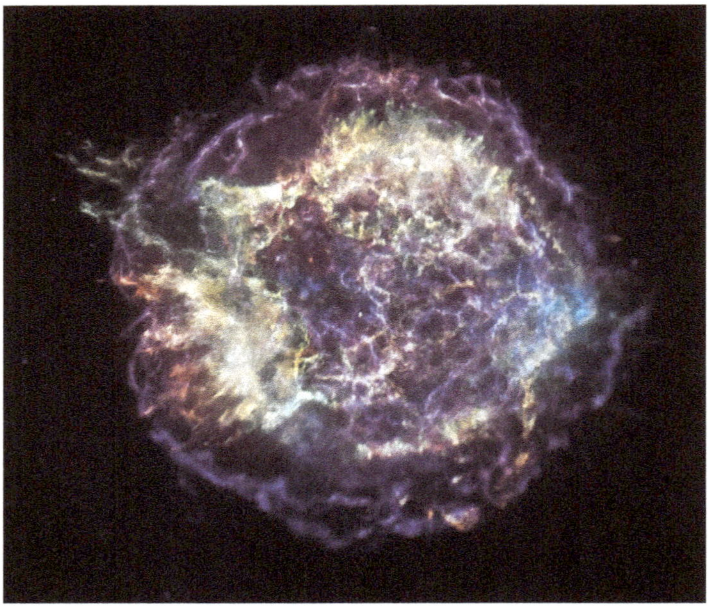

Figure 5.6. Supernova remnant of Cassiopeia A. This picture is a composite of images taken with three classes of observatories: *Spitzer*, which is shown in red and detects infrared radiation; Earth-based telescopes, which detect visible light and shown in yellow; and *Chandra*, which observes X-rays and shown in green and blue. Courtesy of NASA.

antineutrinos (and other neutrinos as well, but it is mostly the electron antineutrinos that will be involved in producing *ee*'s) that cool the star. Neutrinos have a number of interesting properties, so we will discuss them more extensively soon. But for now, just assume that they are another particle. The electron antineutrinos, hereafter denoted as just antineutrinos, will interact differently with the nitrogen nuclei in the amino acids depending on whether the spin of ^{14}N (spin = 1 in units of $\hbar$) is aligned or antialigned with that of antineutrinos (spin 1/2). The relevant reaction has an electron antineutrino interacting with a ^{14}N nucleus to produce a ^{14}C nucleus and a positron, an antielectron:

$$\bar{\nu}_e + {}^{14}\mathrm{N} \rightarrow {}^{14}\mathrm{C} + e^{+}.$$

From basic weak interaction nuclear physics, it is known (Morita 1973) that the cross section for this reaction, if the spins are aligned, is about an order of magnitude less than if the spins are antialigned. This is because in the former case, the conservation of angular momentum requires that one unit of angular momentum comes from the wave function of either $\bar{\nu}_e$ or e^+, and this is what costs the order of magnitude. This, along with the orientation of the ^{14}N spin in the magnetic field, produces a selective destruction of one ^{14}N orientation. Although the *ee*'s in any model of amino acid production in space are usually smaller than some of those observed in meteorites, it is generally accepted that autocatalysis (Gol'danskii &

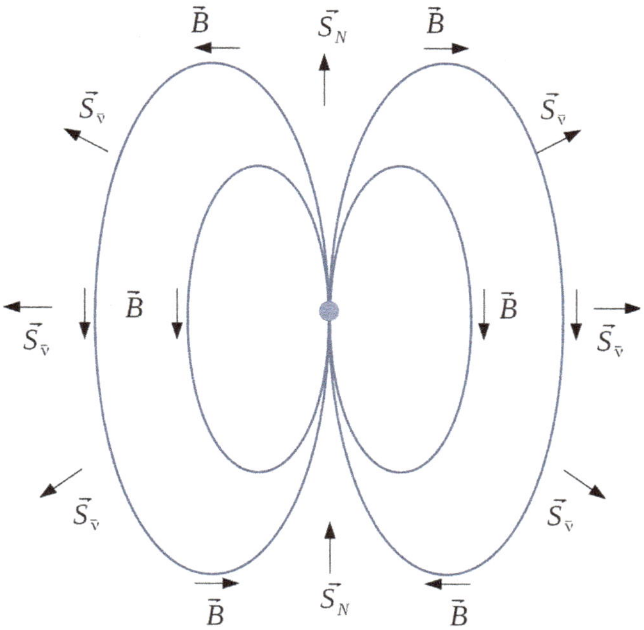

Figure 5.7. Magnetic field lines and spin directions of the antineutrinos and ^{14}N nuclei around a neutron star.

Kuz'min 1989; Kondepudi & Nelson 1985; Mason & Tranter 1985) will drive the small amino acid *ee*'s to the values observed and, ultimately, to Earthly homochirality following their arrival on Earth.

In recent works (Boyd et al. 2010, 2011, 2018; Famiano et al. 2014, 2018a, 2018b), several effects that would couple the N spin to the alanine molecule have been investigated. It was shown that the selective destruction of N would indeed produce an *ee*, which was found to be dominated by the left-handed chirality observed in meteorites. The maximum *ee* found was 0.014%. However, this was the case for isolated alanine. In this book, we describe calculations for amino acids in covalent, zwitterionic, cationic, and anionic geometries. Zwitterionic forms were found to produce considerably larger *ee*'s in most α-amino acids than the other forms. We will describe these calculations in more detail in Chapters 6 and 7.

It is assumed in the SNAAP model that the objects being processed in the magnetic and electric fields generated from the supernova explosion, for example, a passing meteoroid, are large enough that some surface material could be ablated away, leaving a smaller meteoroid. However, it would preserve the *ee* that the larger object would have had, since the small cross section of neutrinos for interacting with matter (Fuller et al. 1999) would allow them to completely process a meteoroid of any size, even one as large as a planet. More ablation would occur later as the meteoroids passed through the atmosphere of whatever planet they ultimately encountered, but the *ee* within the part that finally got to the surface of the planet, with the possible exception of the surface that was heated by passage through the planet's atmosphere (although that would be expected to affect the total number of amino acids and not the *ee*'s), would not be altered.

5.5 Characteristics of the SNAAP Model Star

We will assume as the prototypical SNAAP model star a core-collapse supernova. Its attendant production of electron antineutrinos and intense magnetic field are, as we will show in Chapters 6 and 7, appropriate for inducing molecular effects and orienting the ^{14}N spin to selectively destroy one chiral state.

However, the stars that will ultimately become supernovae have a downside. Stars that have masses of roughly 8–25 times the mass of the Sun (and smaller masses as well) will, in their helium-burning phase of stellar evolution (see Chapter 3), become red giants. After the hydrogen in the core of the star has been consumed and the star begins to burn its helium, the energy that is produced is so enormous that the outer portions of the star are forced out by the pressure of the radiation being produced in the core. The star becomes much larger than it was during its hydrogen-burning phase, and because it has expanded so much, the energy output per square centimeter per second of the surface is reduced, even though the total energy output per second has increased. This will make the surface appear cooler, hence redder, than it was when the star was in its hydrogen-burning phase, thus suggesting the star's red giant name.

The radius of a typical red giant is so much larger than it was when the star was in its hydrogen-burning phase that it will typically extend to a distance that is about the size of the orbit of Earth around the Sun. That will also be well beyond the maximum radius over which the neutron star's magnetic field, produced when the supernova occurs, will be able to provide the alignment necessary for the neutrinos to process molecules. Thus, in order to be processed, the molecules will have to be *inside* the red giant! That, of course, would destroy them, unless they are shielded in some way from the intense heat from the progenitor star.

The maximum distance at which neutrino processing can occur for this single supernova could be determined by the point where the magnetic field from the nascent neutron star or black hole drops to the level of the galactic background magnetic field, about 10^{-6} Gauss—one-millionth of a magnetic field unit (Ferriere 2001; Earth's magnetic field is roughly one-half Gauss and varies across the surface). But this would be an upper limit. When the magnetic field from the neutron star or black hole becomes comparable to this background magnetic field, the direction of the field becomes random, and the selective processing resulting from the alignment due to the magnetic field and neutrino spin can no longer take place. A more stringent limitation is imposed by the dependence of the processing efficiency on the magnetic field intensity; this generally limits the effective processing distance to well inside 1 au (astronomical unit), the distance from the Sun to Earth, or 93 million miles.

What about the stars with masses greater than 25 solar masses? Heger et al. (2003) found that these stars are likely to become what are known as Wolf–Rayet (generally denoted WR) stars, which shed one or two of their outer layers in stellar winds that are the result of the extreme radiation pressure from the cores of these massive stars. Indeed, the products of the nucleosynthesis that occurs in the outer shells can be seen in the resulting clouds: nitrogen, the dominant element resulting from hydrogen burning; and in more massive WR stars, carbon, the result of helium burning (see Chapter 3). Furthermore, the stellar winds are sufficiently thick that they often obscure the central object (Crowther 2007; Williams et al. 1987; Lepine et al. 2000). When astronomers look at most stars with a spectrograph (see the description in Chapter 4), they can determine the intensities of the different constituents of the photosphere, or outer portion, of the star from their absorption lines. But WR stars are different; they produce emission lines that characterize those elements in their winds—the nitrogen or the carbon that the astronomers observe—and completely obscure the emissions from the star itself. That means, however, that the cloud around the WR star has to be hot in order to produce the emission lines.

Another interesting feature is that dust grains have been observed in these clouds (Crowther 2007; Williams et al. 1987). They are apparently made when the wind from a WR star collides with the wind from a massive companion star. We will return to the binary star scenario below. However, this scenario would probably not be able to provide the molecule-containing grains on which our model depends because of the following. First, the grains are thought to occur only beyond several tens of astronomical units (Crowther 2007) from the WR star, and the magnetic field from the nascent neutron star, when it forms, cannot align the molecules beyond a distance somewhat less than 1 au (Boyd et al. 2018). Second, despite the fact that WR stars are quite small (less than the size of the Sun), they are extremely hot, so even though the material in the winds from the star absorbs a lot of the energy from the star, the temperature is still too high, even as far away as 1 au, for grains to form.

By the time a grain gets far enough from the WR star that it could sustain the creation of molecules that would not be destroyed as soon as they formed, there is no magnetic field strong enough to provide the orientation for chiral processing. However, this would not be the case for dust grains or meteoroids that just happened to be passing through the cloud closer to the WR star. There are such objects flitting about the Galaxy all the time. Some of them are within a bit less than 1 au from the star when it becomes a supernova and so will be processed. Indeed, this seems to be the most likely scenario for chirality selection from a single star.

The volume with radius of the distance from Earth to the Sun is huge. During meteorite shower peaks, we Earthlings see a wonderful streak in the sky every minute or so; that is in a volume that is ten trillion times smaller than the volume surrounding a WR star in which the molecules might be processed!

These, of course, would be meteoroids that had been produced somewhere else, so many of them would have had time to produce molecules. These would also be racemic unless they had already been near a supernova when it exploded, in which case the current supernova would just increase the level of enantiomerism.

There is one more thing we need to worry about, that being the high radiation field that the meteoroids and grains will encounter as they pass through enough of the WR star's cloud to get them within range of the magnetic field and the neutrinos from the supernova. This radiation is sufficiently intense that it would completely vaporize small grains, and that would surely destroy any molecules from the surfaces of larger grains.

But this might not be the case for meteoroids that were agglomerations of smaller grains that had some molecules on their surfaces before they got locked into the larger meteoroid. In this case, some of the surface material could be ablated away while still leaving inner material—and enantiomeric molecules. Indeed, the assumption of larger objects passing by may also relax the conditions on the progenitor star—the red giant—to allow supernovae from less massive stars, that is, those that ultimately produce a neutron star. However, it would probably not allow even very large objects to survive the incredible photon blast from a supernova that did not swallow its photons in its subsequent collapse to a black hole, so the WR star may be the only single star entity that will work for the SNAAP model.

5.6 Efficiency Estimate

Can we estimate what the efficiency of chirality selection would be? We can assume that the closest distance that a grain or meteoroid, presumably containing the molecules of interest (created elsewhere), could come to a WR star without having its molecules destroyed is about 10^{12} cm. We can guess what the cross section (the probability for one antineutrino on one ^{14}N nucleus to convert it to ^{14}C) would be from theoretical studies done on other nuclei (Fuller et al. 1999) for antineutrinos undergoing a charged-current weak interaction on ^{14}N; that interaction probability might be around 10^{-40} cm^2. This number is as small as it is for several reasons. First, it requires that the neutrino hit the nucleus, which looks to the incident neutrino like a circle with a radius of about 3×10^{-13} cm (for ^{14}N, that is three ten-thousandths of a billionth of a centimeter), and so has an area of about 3×10^{-25} cm^2. If the incident particle were a sufficiently energetic neutron instead of a neutrino, it would interact via the strong interaction, perform a nuclear reaction, and would have a cross section of about that same size. That is not the case with the neutrino, though, which interacts via the weak interaction; that explains most of the additional factor of 3×10^{-14} in the cross section.

We can assume that there are about 2×10^{56} electron antineutrinos emitted when the supernova explodes, which then gives the probability of any molecule being destroyed by this interaction to be about 2×10^{-9}, two in a billion! This is incredibly small, although it would be larger if we were considering a larger object, for example, a significant meteoroid that happened to wander by as the supernova went off. For this latter object, the distance from the star could be smaller, as the inner part of the meteoroid could be shielded by the outer part. The antineutrinos would, of course, process the entire object. Also, the conversion probability might not be quite as bleak as is suggested by this number. If the object were a factor of 10 closer to the star, its probability of interaction would increase by a factor of 100. In any event, from there, autocatalysis would have to take over to increase the *ee*.

Indeed, it is not at all unreasonable to assume that a meteoroid could be within 10^{11} cm of the star; that is a bit less than 1/100 of an astronomical unit. The meteoroid would have to be initially large enough and be moving sufficiently swiftly so that some of it could be ablated away while protecting the part that remained. It would also have to not be in the high-temperature region of the central star for long enough to have all of its remaining molecules destroyed. If that is the case, now the probability of the molecules in that dust grain or meteoroid being processed are improved: 2×10^{-7}, or one in 5 million. But that is for a single star operating for a short time. Figure 5.8 summarizes the discussions of distance scales that affect our conclusions.

Hence, the supernovae from massive stars can enable the processing that we need to produce enantiomerism in the amino acids if they are included as part of a passing meteoroid, but the level of enantiomerism may still be quite small. Recent calculations, however, assuming different configurations (Famiano et al. 2018; Boyd et al. 2018), have suggested that it is probably considerably larger than the net enantiomerism suggested above. Indeed, it may be large enough that it might not even require autocatalysis to achieve the *ee* levels observed in the meteorites. However, that will require a different scenario than is afforded by a single star.

5.7 The Neutrino Story

Neutrinos have entered our discussions in many places. They are interesting particles in their own right, however, so we want to describe them from a more general perspective than from where they have arisen in our preceding discussions.

We noted in Chapter 1 that neutrinos have no charge, are nearly massless, and interact only through weak interaction. That is not quite true; they also interact through gravitational interaction, but this interaction does not affect the present story and is irrelevant for virtually every situation except those on very large scales, like motions within galaxies and universes, just because neutrinos have such tiny masses.

Neutrinos are in the particle class called leptons, and they come in three types, or flavors: electron neutrinos, mu-neutrinos, and tau-neutrinos, and each has its corresponding antiparticle. Each flavor also has a charged particle: an electron, muon, or tau particle, and each of those has its own antiparticle. Electron neutrinos are produced when a nucleus has too many protons to be stable, as discussed in Chapter 3. This process is called β-decay, in which the nucleus will either capture an electron to convert one of its protons to a neutron or emit a positron, which is an antielectron, again to convert one of its protons to a neutron. These two decay processes look like this:

$$\mathrm{Nuc}(Z, N) + \mathrm{e}^- \rightarrow \mathrm{Nuc}'(Z - 1, N + 1) + \nu_\mathrm{e}$$

or

$$\mathrm{Nuc}(Z, N) \rightarrow \mathrm{Nuc}'(Z - 1, N + 1) + e^+ + \nu_{\mathrm{e}.}$$

In the first reaction, ν_e is an electron neutrino, e^- is an electron, Nuc(Z, N) is a nucleus with Z protons and N neutrons, and Nuc′(Z − 1, N + 1) is a new nucleus with one fewer proton and one more neutron than existed in Nuc(Z, N). The second reaction involves the same nuclei, but now e^+ is a positron. Both processes require

Figure 5.8. Distances relevant to processing grains or meteoroids in the magnetic field and electron antineutrino flux of a core-collapse supernova. The color of the vertical axis gives the relative strength of the magnetic field as a function of distance from the neutron star. The red circle indicates the maximum distance at which any processing could occur, and is roughly 0.01 au. The green line indicates the radius of the Sun. (The bent shape of the lines is because they are projections of circles onto a surface—the magnetic field strength—that is not cylindrically symmetric.)

that the mass energy of Nuc(Z, N) be greater than that of Nuc′(Z − 1, N + 1) + the mass energy of the positron. As discussed in Chapter 3, mass energy here refers to Einstein's famous equation $E = mc^2$ (physicists often express masses in terms of their mass energies).

If that mass energy difference is positive but small, only the first process will occur. If it is greater than about 1.5 MeV, both processes can occur, but the second one will dominate (Boyd 2007). MeV stands for million electron volts, a unit of energy that is especially useful for describing nuclei and nuclear processes, which are often characterized by energies that are on the order of an MeV. As an example, ^{15}O is a radioactive nucleus produced in the hydrogen burning in massive stars (see Chapter 3). It has eight protons and seven neutrons, and decays to ^{15}N (which has seven protons and eight neutrons), a positron, and an electron neutrino. The mass energy of ^{15}O exceeds that of ^{15}N + that of the positron by 2.754 MeV.

Conversely, if a nucleus that has too many neutrons to be stable is formed, it will emit an electron and an electron antineutrino. This is also called β-decay. This occurs as a critical part of either the s- or the r-process, both of which occur in stars to make heavy nuclei; these were discussed in Chapter 3. This β-decay process looks as indicated below, where e^- is an electron and $\bar{\nu}_e$ denotes the electron antineutrino:

$$\mathrm{Nuc}(Z, N) \rightarrow \mathrm{Nuc}''(Z + 1, N - 1) + e^- + \bar{\nu}_e.$$

In both forms of β-decay, a β-particle, that is, an electron or a positron, is involved, either being emitted or captured. An electron antineutrino or neutrino is emitted in both cases in order conserve the lepton number. In special situations, such as those inside neutron stars, a very neutron-rich nucleus can capture a positron, which exists at the high temperatures that exist in such a star, to increase its nuclear charge, and an electron antineutrino will be emitted in the process. This reaction looks like:

$$\text{Nuc}(Z, N) + e^{+} \rightarrow \text{Nuc}'''(Z + 1, N - 1) + \bar{\nu}_e.$$

We will come back to that reaction in Chapter 7.

At the temperatures, hence energies, that exist in the Sun, only electron neutrinos and electron antineutrinos could be produced, but actually, only electron neutrinos are emitted. That is due to the nuclear reactions that take place in the Sun, as explained in Chapter 3.

All neutrinos and antineutrinos are spin 1/2 particles, but the neutrinos and antineutrinos have opposite chirality. The ones we will be concerned with are the electron neutrinos and antineutrinos; the spin of the electron antineutrino will be parallel to its direction of motion, whereas that of the electron neutrino will be antiparallel to its direction of motion. This will play a crucial role in the selection of molecular chirality, as we will see below.

However, at higher energies, for example, those that arise when high-energy cosmic rays hit Earth's atmosphere, pions can be produced. Pions are unstable particles that can be generated when the energies of colliding particles exceed the value of the pion mass energy, $m_\pi c^2$, 140 MeV. The pion quickly decays into a muon and the appropriate mu-neutrino, as indicated below. The symbols π^+ or π^- denote a positively or negatively charged pion (there can also be a pion with zero charge, but it decays quite differently), μ^+ or μ^- denote a positively or negatively charged muon, and ν_μ and $\bar{\nu}_\mu$ denote, respectively, a muon neutrino and a muon antineutrino:

$$\pi^+ \rightarrow \mu^+ + \nu_\mu$$

or

$$\pi^- \rightarrow \mu^- + \bar{\nu}_\mu.$$

When the muon decays, it decays into an electron if it is a μ^- or to a positron if it is a μ^+, and a neutrino and an antineutrino. These decays look like:

$$\mu^+ \rightarrow e^+ + \bar{\nu}_\mu + \nu_e$$

or

$$\mu^- \rightarrow e^- + \nu_\mu + \bar{\nu}_e.$$

In each of these reactions, the lepton number is conserved, that is, the number of leptons (the electrons, muons, neutrinos, and their antiparticles) on the left-hand side of the equation is the same as their number on the right-hand side. A lepton counts as +1 particle, and an anti-lepton as −1. So, for example, in the last equation, the μ^- on the left-hand side counts as +1 lepton, as does the e^- on the right-hand side.

The ν_μ counts as +1 lepton, while the $\bar{\nu}_e$ counts as −1. So everything adds up as it should.

To complete the story, tau particles can be produced in still higher energy interactions. They have such a short lifetime that they barely move at all, unless they are highly relativistic, in which case their decay time will be lengthened by time dilation, an effect that exists for particles moving near the speed of light. However, when they decay, they will produce tau-neutrinos among their decay products, just as the muons produced mu-neutrinos.

5.8 Interactions of Neutrinos with ^{14}N

At the lower energies that are relevant to the effects of neutrinos and antineutrinos on amino acids, we only need to be concerned about the electron antineutrinos and electron neutrinos that can change ^{14}N to either ^{14}C or to ^{14}O, respectively. In both cases, we are relying on charged-current weak interactions to mediate the process. We include that to discriminate between that interaction and neutral-current weak interactions, which would produce different effects that would not be chirally selective.

In addition, we note that the cooling neutron star that was produced from the supernova can emit not only electron neutrinos and antineutrinos, but mu- and tau-neutrinos and antineutrinos. Those neutrinos can be produced deep inside the star as neutrino–antineutrino pairs, a process that requires much less energy—only $2m_\pi c^2$, and the neutrino masses are all really tiny—than necessary to produce a muon or a tau particle via a charged-current weak interaction.

Although the energies of neutrinos from a supernova can get to be quite high, they do not come anywhere near the thresholds imposed by the requirement that the energies must exceed the rest mass energy of muons—105 MeV in rest mass energy units. (The electron mass in those same units is 0.511 MeV, or for tau particles, it is 1777 MeV.) The electron neutrinos and antineutrinos operate on ^{14}N through the following reactions:

$$^{14}\mathrm{N} + \bar{\nu}_e \rightarrow {}^{14}\mathrm{C} + e^+$$
$$^{14}\mathrm{N} + \nu_e \rightarrow {}^{14}\mathrm{O} + e^-.$$

In the first reaction, the electron antineutrino $\bar{\nu}_e$ converts ^{14}N to ^{14}C, and in the second, the electron neutrino ν_e converts ^{14}N to ^{14}O. Note that these reactions look just like the β-decays that we discussed in Chapter 3 in the context of hydrogen burning in massive stars, except that here the neutrino appears on the left side of the equation instead of on the right side. However, these reactions are mediated by the weak interaction, just as β-decay. Also, if you switch a neutrino, antineutrino, electron, or positron from one side of the equation to the other, you need to change it to its antiparticle (to obey the lepton conservation law). The neutrino energies required to perform these reactions are important, as the reaction probabilities depend on the energy.

The actual neutrino energies from a supernova are uncertain. However, the studies of Georg Raffelt and his group (Esteben-Pretel et al. 2007; Keil et al. 2003), as well as George Fuller and his group (Duan et al. 2006, 2007a, 2007b), suggest that so much mixing of the neutrino flavors, or types, will take place in the stars that all six of the neutrino and antineutrino flavors will emerge from the star with similar energy spectra.

One prediction of the neutrino energies, from Yüksel & Beacom (2007), is shown in Figure 5.9. Although it shows what the expected energies of the electron antineutrinos will be, it is roughly representative of all the neutrino types. Two different neutrino interaction scenarios were tried; the peak energy was found to be around 12 MeV in both cases.

This energy is important because the threshold energy for the $^{14}N + \bar{\nu}_e \rightarrow {}^{14}C + e^+$ reaction is 1.18 MeV (that is the energy that the antineutrino is required to have for the reaction to take place), while the threshold energy for the $^{14}N + \nu_e \rightarrow {}^{14}O + e^-$ reaction is 5.14 MeV. Thus, both reactions are endothermic; energy is required for them to occur. That energy has to be supplied by the incident antineutrino or neutrino. This means that many more electron antineutrinos will be able to convert ^{14}N to ^{14}C than electron neutrinos will be able to convert ^{14}N to ^{14}O.

In addition, most models of neutrinos emitted from supernovae have electron neutrinos somewhat less energetic, by roughly 2 MeV, than electron antineutrinos, and this will further enhance the preference for the ^{14}N to ^{14}C conversion. Finally, the reaction probabilities increase with the amount of energy the neutrino has in excess of the threshold energy; this will further enhance the importance of

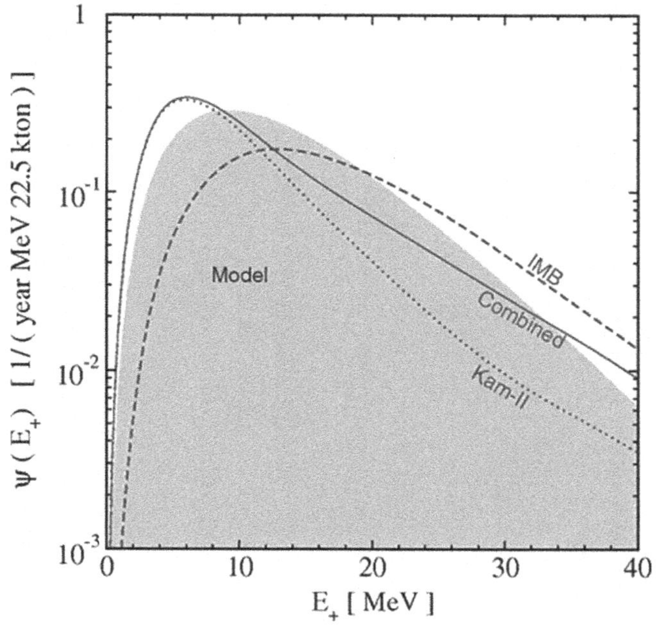

Figure 5.9. Energy spectrum from the neutrinos detected from SN 1987a, as determined from a theoretical analysis of those observed in two detectors, the Kamiokande II (Japanese) detector and the Irvine–Michigan–Brookhaven (United States) detector. Reprinted from Yüksel & Beacom, 2007, Phys. Rev. D, 76, 083007. Copyright 2007 by the American Physical Society. doi.org/10.1103/PhysRevD.76.083007.

the reaction that converts ^{14}N to ^{14}C compared to that which converts ^{14}N to ^{14}O. Suffice it to say that, given the expected energies of the neutrinos from a supernova, the conversion to ^{14}C is considerably more probable than the conversion to ^{14}O. Thus, we will ignore the conversion to ^{14}O for now.

The nuclear physics of what happens is crucial to the success of the SNAAP model. It was summarized earlier in this chapter and is as follows: when the electron antineutrino and ^{14}N spins line up, it is more difficult for the antineutrino to do the conversion to ^{14}C than if they are not aligned. This is related to the chirality of the molecules, as we will discuss below, so this will produce a selective destruction of the ^{14}N nuclei, and hence of the molecules they were in, which depends on the relative nuclear and neutrino spin orientations.

The details involve some simple vector addition. The spin of the neutrinos is 1/2 (in units of $\hbar$, of course), so when the spin of the antineutrino and the spin 1 ^{14}N nuclei are aligned, the total spin is 3/2. When they are antialigned, it is 1/2. The rules of quantum mechanics only allow integer or half-integer spins and angular momenta.

The laws of physics require that the total angular momentum, including spin, be conserved in any reaction, that is, the sum of the angular momentum vectors before an interaction or a decay must be the same as their sum afterward. What happens is shown in Figure 5.10. The positron on the right-hand side of the equation above has a spin of 1/2, so if the total spin was 1/2, as for the antialigned case, the total spin before the reaction is equal to the total spin after the reaction, since the final nucleus, ^{14}C, has zero spin, as indicated in Figure 5.10. But in the aligned case, that will not be the situation unless we come up with one additional unit of angular momentum from somewhere. That will have to be supplied by the electron antineutrino or the positron; it will have to come from one of their quantum mechanical wave functions.

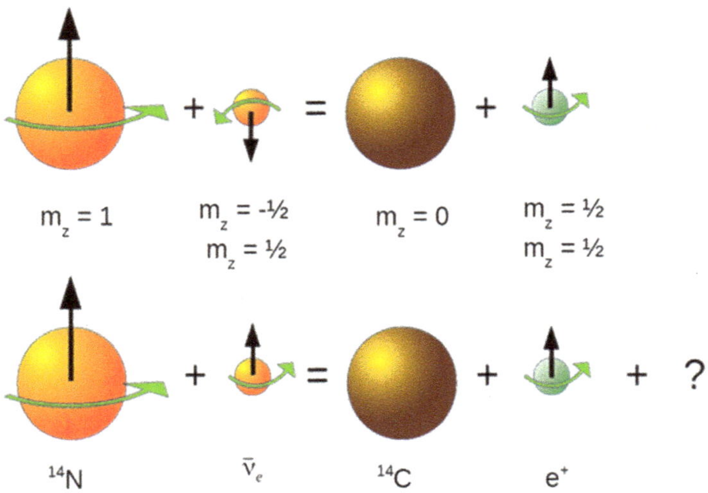

Figure 5.10. Spins of nuclei and particles relevant to SNAAP model chirality selection. See the text for an explanation of the effect of the vector additions.

Again, these reactions are mediated by the weak interaction, and that interaction allows the antineutrino or positron to provide the one unit of angular momentum that is required here. This interaction is the same one that mediates nuclear β-decay, and that has been studied in great detail. We know from those studies that the requirement of angular momentum conservation leads to successive levels of forbiddenness in β-decay, and it will have the same effect on the neutrino-induced interactions. Thus, these reactions are much less probable, by roughly a factor of 10, as mentioned previously, if even one unit of angular momentum needs to be provided by the antineutrino or positron. (If two units needed to be provided, the inhibition would be roughly a factor of 100.) Thus, the reaction will be much more likely for the antialigned case.

Supplying one unit of angular momentum also introduces a change of parity. But both ^{14}C and ^{14}N ground states have positive parity. Thus two units of angular momentum are needed to allow the transition and accommodate parity conservation. However, in our calculations of enantiomeric excess, to be described in Chapter 7, we have adopted the (conservative) assumption of ten for the inhibition for the aligned angular momentum case.

In every location, the ^{14}N spin vectors and the neutrino spin vectors are not just aligned or antialigned, but are at some more complicated angle. Thus, if one is going to try to calculate the enantiomeric excess that will result from the electron antineutrino–^{14}N interactions, it might get pretty complicated. We will describe how to do that in the next two chapters.

Earlier, we dismissed the interactions in which the electron neutrinos would convert ^{14}N to ^{14}O because the reaction threshold was about 4 MeV higher for that reaction than for the one in which ^{14}N is converted to ^{14}C by electron antineutrinos. But the former reaction will occur and will produce exactly the opposite effect that the electron antineutrinos do. It just will not be as frequent as the reaction in which the antineutrinos convert ^{14}N to ^{14}C, so the net result will be the chiral selection by the electron antineutrinos when one averages over all space. And since the conversion to ^{14}O by the neutrinos occurs in the same region of space as the conversion to ^{14}C by the antineutrinos, the latter conversion will dominate. So, our dismissal of neutrino conversion to ^{14}O was justified.

There is one more aspect of core-collapse supernovae that is worth noting. We have already noted that supernovae can outshine all the other stars in its galaxy for a while. This is because they produce a lot of ^{56}Ni, which is radioactive and which decays to ^{56}Co, and then to ^{56}Fe. These decays produce photons, which heat their environment and cause it to shine. All these take place with a time constant of about 80 days (the half-life of ^{56}Co is 77 days), so the light emitted from these supernovae usually falls off on that timescale (see Figure 5.11, which also shows some exceptions to the 80 day falloff). However, note that not all core-collapse supernovae emit the same amount of light (unlike Type Ia supernovae, which we discussed in Chapter 2 as serving as the standard candles for cosmological distance determination).

All of the stars shown in Figure 5.11 are shown as absolute magnitudes, that is, the amount of light one would see if one were at the same distance from each of them. Astronomers must have a system that allows them to discuss and graph sky

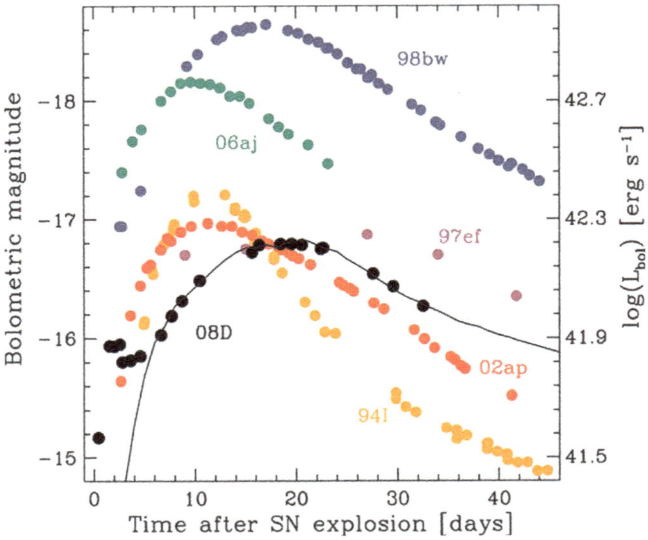

Figure 5.11. Light curves from core-collapse supernovae. Actually, these are all Type 1b/c supernovae, stars that have lost one or two of their outer shells and became Wolf–Rayet stars before they became supernovae. Reprinted from Mazzali et al. 2008, Science, 321, 1185. Courtesy of Science.

objects that have luminosities that vary over many orders of magnitude. Thus, they have adopted a system of magnitudes. The *y*-axis in Figure 5.11 is sort of logarithmic; one magnitude amounts to a difference in luminosity of a factor of 2.512, two magnitudes to a difference of $(2.512)^2$, etc. (If you are not an astronomer, this system seems rather arcane. It originated with the ancient Greeks, but was formalized in 1856 by Norman Robert Pogson, who defined a first-magnitude star as one that is 100 times as bright as a sixth-magnitude star. This implies that a star of magnitude m is about 2.512 times as bright as a star of magnitude $(m + 1)$.)

SN 2008fz was described in the papers that were published about it as the brightest supernova ever observed, so it is at one extreme of the luminosity range of core-collapse supernovae. At the other extreme are the silent supernovae, which may not be seen at all, at least in visible light. The interactions that neutrinos have with ^{14}N nuclei are pretty feeble, and in any case are much smaller than the interactions that photons would have. So, it is reasonable to ask whether the photons would not destroy all of the molecules on which the neutrinos had performed chiral selection. For many supernovae, that would probably be the case.

As discussed above, silent supernovae collapse to black holes before very many photons can be emitted, but they emit all of the neutrinos that supernovae that produce neutron stars do. Heger et al. (2003) studied the various modes of core-collapse supernovae that can occur. They found that core-collapse supernovae occurring in stars having masses less than 25 times the mass of the Sun would ultimately produce a neutron star remnant. However, those from stars ranging from 25 to 40 times the mass of the Sun will first form a neutron star, but, as noted previously, the infalling matter would cause the neutron star cores to exceed the maximum mass that their degeneracy pressure can withstand, resulting in fallback black holes. A star more massive than 40 times the Sun's mass would be expected to collapse directly to a black hole.

In both the fallback black hole scenario and the direct collapse to a black hole, the timescale is so short that much of the matter close to the black hole would be swallowed by the black hole, taking most of the electromagnetic radiation produced in the initial explosion with it. So, the supernova would establish its magnetic field and emit its neutrinos, but would spare the surrounding regions of its burst of electromagnetic radiation. Since ^{56}Ni would be produced deep in the stellar core prior to the supernova, it would also go into the black hole (Fryer 2009). Thus, most of the black-hole-forming stars would preserve the molecules on which the magnetic field and the neutrinos had performed their chiral selection. These supernovae could produce the chirality selection of amino acids that would ultimately seed the Galaxy, or at least the part of the Galaxy closest to the supernova, with a preferred chirality. However, they are not the only potential system for producing enantiomeric excesses in the amino acids. We will return to the progenitor stars in Chapters 6 and 7.

References

Avalos, M., Babiano, R., Cintas, P., et al. 1998, ChRv, 98, 2391
Bada, J. L., Cronin, J. R., Ho, M. S., et al. 1983, Natur, 301, 494
Berman, H. M., Westbrook, J., Zukang, F., et al. 2000, Nucleic Acids Res., 28, 235
Bonner, W. 1991, OLEB, 21, 59
Bonner, W. 2000, Chirality, 12, 114
Boyd, R. N. 2007, An Introduction to Nuclear Astrophysics (Chicago, IL: Univ. Chicago Press)
Boyd, R. N., Famiano, M. A., Onaka, T., Kajino, T. & Mo, Y. 2018, ApJ, 856, 26
Boyd, R. N., Kajino, T. & Onaka, T. 2010, AsBio, 10, 561
Boyd, R. N., Kajino, T. & Onaka, T. 2011, Int. J. Mol. Sci., 12, 3432
Buckingham, A. D. 2004, CPL, 398, 1
Buckingham, A. D. & Fischer, P. 2006, CP, 324, 111
Cronin, J. R. & Pizzarello, S. 1997, Sci, 275, 951
Crowther, P. A. 2007, ARA&A, 45, 177
Duan, H, Fuller, G M, Carlson, J & Qian, Y-Z 2006, PhRvD, 74, 105014
Duan, H., Fuller, G. M., Carlson, J. & Qian, Y.-Z. 2007a, PhRvD, 75, 125005
Duan, H., Fuller, G. M. & Qian, Y.-Z 2007b, PhRvD, 76, 085013
Engel, M. H. & Nagy, B. 1982, Natur, 296, 837
Esteben-Pretel, A., Pastor, S., Tomas, R., Raffelt, G. G. & Sigl, G. 2007, PhRvD, 767, 125018
Famiano, M. A., Boyd, R. N., Kajino, T., et al. 2014, Symmetry, 6, 909
Famiano, M. A., Boyd, R. N., Kajino, T. & Onaka, T. 2018a, AsBio, 18, 190
Famiano, M. A., Boyd, R. N., Kajino, T., Onaka, T. & Mo, Y. 2018b, Scientific Reports, 8, 8833
Faraday, M. 1846, RSPT, 136, 1
Ferriere, K. M. 2001, RvMP, 73, 1031
Fryer,, C. 2009, ApJ, 699, 409
Fuller, G. M., Haxton, W. C. & McLaughlin, G. C. 1999, PhRvD, 59, 085005
Gol'danskii, V. I. & Kuz'min, V. V. 1989, SvPhU, 32, 1

Guijarro, A. & Yus, M. 2009, The Origin of Chirality in the Molecules of Life (Cambridge: RSC Publishing)
Haldane, J. B. S. 1960, Natur, 185, 87
Heger, A., Fryer, C. L., Woosley, S. E., Langer, N. & Hartmann, D. H. 2003, ApJ, 591, 288
Johnson, A. P., Cleaves, H. J., Dworkin, J. P., et al. 2008, Sci, 322, 404
Keil, M. Th., Raffelt, G. G. & Janka, H. T. 2003, ApJ, 590, 971
Kennedy, D. & Norman, C. 2005, Sci, 309, 75
Kondepudi, D. K. & Nelson, G. W. 1985, Natur, 314, 438
Kvenvolden, K., Lawless, J., Pering, K., et al. 1970, Natur, 228, 923
Lepine, S., Moffat, A. F. J., St-Louis, N., et al. 2000, AJ, 120, 3201
Mason, S. F. & Tranter, G. E. 1985, RSPSA, 397, 45
Mazzali, P. A., Valenti, S., Valle, M. D., et al. 2008, Sci, 321, 1185
Meierhenrich, U. J. 2008, Amino Acids in Chemistry, Life Sciences, and Biotechnology (Heidelberg: Springer)
Miller, S. L. 1953, Sci, 117, 528
Miller, S. L. & Urey, H. C. 1959, Sci, 130, 245
Morita, M. 1973, Beta Decay and Muon Capture (Reading, MA: W. A. Benjamin)
Oparin, A. I. 1957, The Origin of Life on Earth (tr. A. Synge; London: Oliver and Boyd)
Pasteur, L. 1948, Researches on the Molecular Asymmetry of Natural Organic Products (Edinburgh: E. and S. Livingstone Ltd.)
Rode, B. M., Fitz, D. & Jakschitz, T. 2007, Chem. Biodiversity, 4, 2674
Ulbricht, T. L. V. & Vester, F. 1962, Tetrahedron, 18, 629
Vester, F., Ulbright, T. L. V. & Krauch, H. 1959, NW, 46, 68
Williams, P. M., Van der Hucht, K. A. & The, P. S. 1987, A&A, 182, 91
Yüksel, H. & Beacom, J. F. 2007, PhRvD, 76, 083007

Richard N Boyd and Michael A Famiano

Chapter 6

Determining Molecular Properties by Quantum Molecular Calculations

6.1 Recap

Previous chapters have indicated what might happen when meteoroids carrying amino acids find themselves in the presence of external electric and magnetic fields. Chapter 5 discussed neutrinos and the interactions they might undergo with nitrogen nuclei. The combination of these factors can be shown to select one amino acid chirality over the other. Amino acids have been found in many meteorites, so we know they are made in space. It also seems reasonable to assume that they are initially created with equal amounts of L- and D-enantiomers. Thus, if a cosmic site with the magnetic field necessary to create chiral dependence can be found, along with the antineutrinos to process the molecules, it should provide an enantiomeric excess.

6.2 Some Background: Magnetic Fields and Nuclei

Atomic nuclei with spin (even neutrons) have magnetic moments and therefore will align themselves in magnetic fields. If a handful of ^{14}N nuclei is placed in a magnetic field, they will all tend to align the z-component of their spins parallel to the magnetic field (assumed to define the z-direction). While coordinate systems are usually arbitrary, the magnetic field provides a convenient choice for the direction of the z-axis. The spin also has x- and y-components. These components precess about the z-axis in a magnetic field, producing a net alignment of the nitrogen nuclei with the magnetic field, giving them a net nuclear magnetization.

Atomic nuclei have charge and currents, and these are distributed in complicated ways throughout the nucleus, subject to the rules of quantum mechanics. Even though a ^{14}N nucleus has a spin of 1, this spin is determined (as it is for all nuclei) by the sum of the spins of all nucleons in the nucleus. Many nucleons pair up with their spins pointing in opposite directions to get a net spin of zero for the pair. But if there is one neutron or proton that has not paired up with another neutron or proton, the

doi:10.1088/978-0-7503-1993-5ch6

total spin of that nucleus is just the spin of the lone unpaired nucleon. In the case of ^{14}N, there is one unpaired proton and one unpaired neutron. Together, they add up to give nitrogen its spin.

The magnetic moment for any nucleus is proportional to its spin. As indicated in Chapter 5, the magnetic moment and the nuclear spin are related through the gyromagnetic ratio, the ratio of the nuclear magnetic moment to the nuclear angular momentum.

The magnetization vector for some solid object is the net number of magnetic moments per unit volume pointing in a particular direction times the magnetic moment of each particle in the object. For an amorphous material that is not in any external magnetic field, the magnetic moments of the individual particles point in random directions, and the magnetization is equal to zero. For a permanent magnet, with magnetic domains, there is a net magnetization, meaning that many dipole moments are constrained to point in the same direction.

If the solid material is in an external magnetic field, then that field may cause the individual magnetic moments to align with or against the field. The induced magnetization of a material in an external magnetic field relative to the external field is called its magnetic susceptibility. Not everything has a magnetization until you put it in a magnetic field; many things have induced magnetizations. In some cases, a substance will have a small induced magnetic moment that aligns with the field. These types of materials are called paramagnetic. Such materials are weakly attracted by an external magnetic field, and they do not retain their magnetism when the magnetic field is removed. Some materials—like iron—are ferromagnetic. They have a large, positive magnetic susceptibility. Some materials are diamagnetic. They have a negative susceptibility. Putting them in a magnetic field will actually induce magnetic moments in them that point opposite to the direction of the external field. They are slightly repelled by the external field. These materials have paired electrons that have no permanent magnetization. In an external field, these electrons tend to move in such a way as to oppose the external field. This is a direct result of a physical principle known as Lenz's Law. Most elements in the periodic table, including some metals, are diamagnetic.

However, for atomic nuclei, there are no interfering electrons, not directly anyway (more on that later). Thus, while we might expect electron orbitals to have induced magnetic moments (the magnetic susceptibility), we expect atomic nuclei to align with the external field (ignoring any temperature-dependent effects, which we will discuss later). We would expect the magnetization for the atomic nuclei only to be equal to the total number of nuclei times their magnetic moments.

6.3 Nuclear Magnetic Resonance in Space

To consider the possible effect of nuclear magnetization for atomic nuclei bound in meteoroids located in external magnetic fields, one has to model a population of nuclei. They may be frozen in ice that has collected inside the meteoroid. As this meteoroid moves through space, it might pass through a very strong magnetic field. The meteoroid is, of course, spinning randomly with respect to the field. Consider

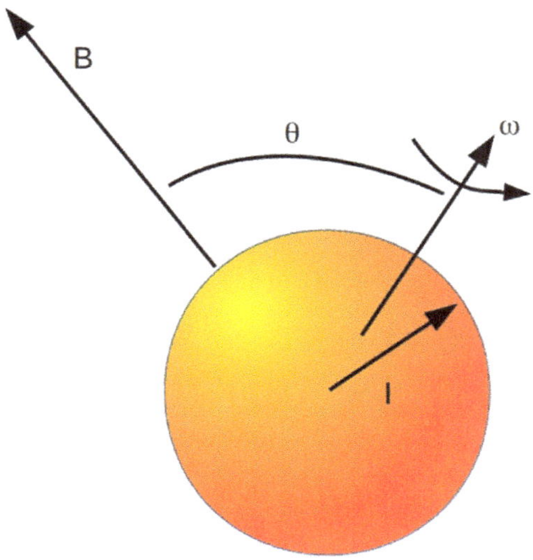

Figure 6.1. The vectors relevant to the dynamical model of chiral state selection in the SNAAP model. ***B*** is the magnetic field vector, ***I*** is the total spin of the molecule, and ***ω*** is the vector that characterizes the rotation of the meteoroid containing the molecules.

the several vectors shown in Figure 6.1 (Famiano et al. 2014; Famiano & Boyd 2016). These include the magnetic field vector ***B***, the total nuclear angular momentum vector ***I***, and the vector ***ω*** that characterizes the rotation of the meteoroid containing the molecules. The random orientation of any particular meteoroid with respect to the magnetic field can be specified by an angle θ.

Consider what happens in the reference frame of the meteoroid. For a material consisting of nonzero magnetic moments in an external magnetic field ***B***, the individual nuclear angular momentum vectors will precess about the magnetic field at a rate that is a function of the magnetic moments of the nuclei. In the meteoroid's reference frame, the magnetic field precesses about the vector ***ω***. One can then think of the magnetic field as having two components in the meteoroid's reference frame: one parallel to ***ω***, which we will call $\boldsymbol{B}_0$, and one perpendicular to ***ω***, which we will call $\boldsymbol{B}_1$. Their values will depend on ***B*** and θ. The parallel component, $\boldsymbol{B}_0$, is stationary in the rest frame of the meteoroid, and $\boldsymbol{B}_1$ rotates around ***ω*** with the frequency at which the meteoroid spins.

Thus, this model has several moving parts. The meteoroid is spinning. The nitrogen nuclear spins are precessing about the magnetic field axis, and $\boldsymbol{B}_1$ is rotating about the rotation axis of the meteoroid. In the rest frame of the meteoroid, the nitrogen nuclei are precessing about $\boldsymbol{B}_0$ while $\boldsymbol{B}_1$ rotates about $\boldsymbol{B}_0$ independently, nudging the nitrogen nuclei slightly off-axis. Depending on the angle between ***B*** and ***ω***, this may be a big push or a slight nudge. It could be large enough to completely disorient the spins of the nitrogen nuclei.

However, if $\boldsymbol{B}_1$ rotates around $\boldsymbol{B}_0$ at the same rate at which the nuclei precess about the magnetic field, then every time the nitrogen nuclei are at the same spot in their precession, they are given a small push by $\boldsymbol{B}_1$. It will not take too many rotations before a significant portion of the nitrogen nuclei are all pointing in the same direction under the influence of the external magnetic field, and we measure an increase in the magnetization of the nuclei. They are then said to be polarized. This is known as a resonant condition. These days, external radio-frequency pulses are commonly used to flip the magnetization of the nuclei.

The precession frequency of ^{14}N in a magnetic field can be quite high, and it is hard to imagine meteoroids rotating that fast. However, there are two things that might affect this situation. One is that the strength of the field can be lower so that the precession frequency is smaller. Another is to keep in mind that the oscillation of $\boldsymbol{B}_1$ does not have to be at the precession frequency. It can be an integral multiple of the precession frequency. The nudge can be applied every 2nd, 3rd, 4th, 10th, 100th, etc., lap. Of course, there are other factors that come into play, which we will discuss later.

This same effect is exploited in nuclear magnetic resonance (NMR). A magnetic polarization is created in a material. This creates a signal that can be picked up. Physicians exploit this property in magnetic resonance imaging (MRI) to "see" inside people. Different tissues have different resonant frequencies, and these will create signals that can be pinpointed in position to create an image. We can polarize the medium using this technique (Bloch 1946).

NMR experiments in the laboratory are much more controlled than NMR-like effects for meteoroids rotating randomly in space. To examine the outcome of this effect, we performed several Monte Carlo simulations of meteoroids in the vicinity of a neutron star's magnetic field to simulate a spatial distribution of meteoroids with varying angular speeds and orientations relative to the external dipole field (Famiano et al. 2014). A Monte Carlo simulation varies the parameters of the object under study according to some assumed probability distribution. Parameters are randomly chosen to fit this distribution, and physical quantities (which we know how to calculate) are computed based on these parameters. In the case of this meteoroid model, the random parameters are the meteoroid rotation rate and the angle of the meteoroid with respect to the magnetic field. Once these are known, the polarization using NMR formulae can be computed.

For these Monte Carlo simulations, meteoroids were assumed to be distributed evenly in a volume of space about the neutron star with random angular orientations and velocities. This means that randomly distributed distances from a neutron star, which was the source of the magnetic field, were chosen. Random orientations and speeds were also assumed. For each sample in the Monte Carlo calculation, the components of the magnetization were calculated along with the polarization angle.

Some results of the simulations are shown in Figure 6.2. Three histograms of the polarization angle are presented for three different conditions. The polarization angle is the angle of the bulk magnetization with respect to the external magnetic field. It can be seen that it is possible to have conditions in which the nuclei have no preferred polarization and situations in which they are quite polarized. In this figure, the angular speed of the meteoroid in each event varies uniformly between 0 and 10 rad s^{-1}, and the meteoroids were assumed to be situated at distances from the

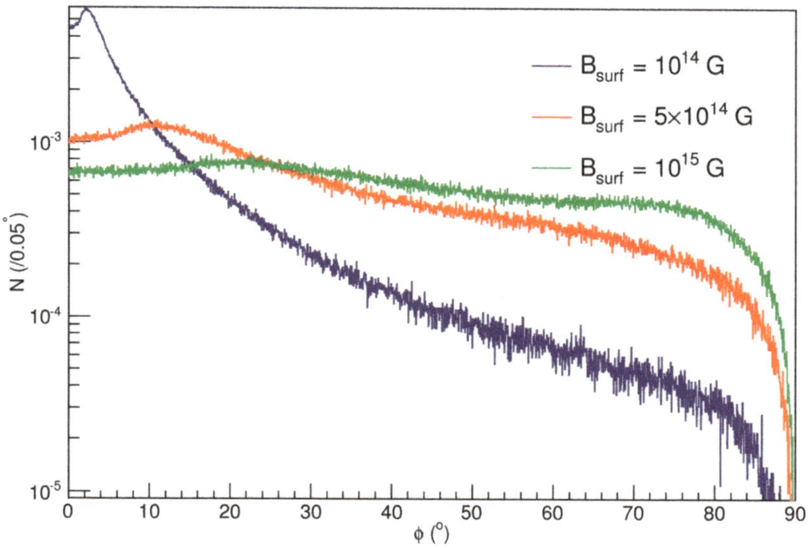

Figure 6.2. Polarization distribution for the three models in the Monte Carlo calculations. This figure compares the polarization distribution for various neutron star surface magnetic fields.

neutron star between 0.01 and 0.05 au. It is apparently possible under some conditions to produce some nuclear alignment. All of these assume that the nitrogen nuclei stay where they are in the meteoroid. For some conditions, this may be a good approximation, as some nuclei in some molecules may hold their polarization in cold environments for several minutes or more.

However, the nitrogen nuclei do not stay polarized unless they are at zero temperature. The effects of temperature on magnetic polarization need to be considered.

6.4 Thermal Effects

Putting a population of atomic nuclei in a magnetic field will cause their magnetic moments to tend to align with the field. However, not all of the nuclei align with the external magnetic field. Thermal motion will cause some nuclei to align with the field and some against it. In the case of the spin-1 ^{14}N nucleus, there are three magnetic substates, with quantum numbers −1, 0, and +1. The spin-0 state aligns perpendicular to the field. We can also describe the energy states of each nucleus. When the magnetic moments are all aligned with the field, this is the lowest energy state of the nuclei.

If the meteoroid's environment is at nonzero temperature, several things can happen. Photons will collide with atomic nuclei, possibly imparting enough energy to them to change some spin states to a higher energy state, in which the spins of some nuclei are antiparallel to the magnetic field. An even higher temperature will increase the number of photons as well as their energy distribution, and promote more nucleons into higher energy states. At some point, the entire mix will be evenly distributed in the three spin states, and they will be continuously trading places. The temperature is closely related to the entropy of the system, which is a measure of how chaotic the system's states are and which is related to how many different ways there are of arranging the spin states of a system. If all of the spins point in one direction, the system has the lowest possible entropy; there is only one way of arranging all of the spins to get that state.

However, if the temperature is low enough and the field high enough, we will have a system with a significant number of spins aligned parallel to the field. Thus, the temperature tends to counter the effects of the external field in aligning the nuclei.

6.5 Electric Fields and Molecules

The effect of magnetic fields will turn out to be important in achieving amino acid chirality: the orientation of the ^{14}N nuclei within the magnetic field provides a basis for selecting one molecular chirality over the other. Putting the nucleus in the field provides a population of nuclei with their spins at least partially aligned with the direction of the field. The field also has another effect that will turn out to be important. Many molecules are diamagnetic; they do not have large magnetic moments. That is not to say that the molecules are not affected by the magnetic field, however. In fact, the presence of molecular electrons can change the field slightly at the location of the nucleus. On the other hand, because molecules generally have no magnetic moment, they are not strongly aligned in the external field. All this is to say that atomic nuclei are strongly affected by external magnetic fields, while the molecules they are in are not so strongly affected. The nucleus may reorient itself within the molecule, but the molecule will feel little to no effect from the external magnetic field.

This makes it possible for the nuclei in amino acids to affect their chirality, which is a *molecular* property. To demonstrate this, two more pieces need to be discussed:

- The entire molecule must be oriented with respect to the nuclear spin.
- There must be a difference in the molecular and nuclear orientations between the two possible chiral states.

While molecules may not necessarily have strong magnetic dipole moments, they may have considerable electric dipole moments. The molecular magnetic dipole moment is due to the distribution of current in a molecule (electron velocity distribution), but the electric dipole moment is just due to the distribution of charge (electron position distribution). Since molecules have electrons orbiting at considerable distances from their nuclei, these electrons can create a charge distribution that covers a large spatial extent within the molecule. Furthermore, the charge distribution very much depends on the molecular shape. Consider, for example, the water molecule, represented in Figure 6.3 in two different ways. The image of water with an oxygen sphere attached to two hydrogen spheres, depicted on the left, is more familiar to many people. However, the electron distribution for water, right image, is what gives it its electric dipole moment. Oxygen's larger nuclear charge tends to make the electrons cluster around it rather than around the hydrogen nuclei, creating a larger electron density near the oxygen nucleus and giving the molecule a charge distribution that is definitely not spherical.

Figure 6.3. Two different representations of the water molecule. On the left side is the familiar ball-and-stick model with the oxygen atom in red and the hydrogen atoms in white. On the right side is a surface figure, which is more representative of the electron distribution (Berman et al. 2000). The surface is one of constant charge density. (The colors represent results from an electron spin polarization calculation. In the figure, red represents the highest density and blue the lowest.)

If a water molecule is placed in an external electric field, the electric field exerts a torque on it in the same way that a magnetic field exerts a torque on a nucleus with a magnetic dipole moment. The electric dipole moments exhibit energy and thermal behavior similar to magnetic dipole moments in magnetic fields. That is, the electric dipole moment aligned with the electric field is the lowest energy state, and they become less aligned with increasing temperature.

Atomic nuclei, on the other hand, generally have very small electric dipole moments. Thus, the nuclei are oriented strongly by a strong external magnetic field, while the molecules are oriented by an external electric field. This becomes more interesting when we consider chiral molecules. Because the molecular shape is determined by the location of the electron orbitals about the nuclei, the electric dipole moment of the molecule depends on the chirality. It points in different directions for the L- and D-enantiomers. This is shown for the amino acid alanine in Figure 6.4. Since the electric dipole moment depends on the molecular geometry, it is also a mirror image. If we had an L- and a D-enantiomer in an external electric field, their lowest energy states would be the ones in which the electric dipole moment aligned with the field. We can see that the enantiomers would have different orientations in an external electric field.

Thus, if we have both electric and magnetic fields, we can get a population of nuclei and molecules coupled to different fields and aligned in specific directions. This will be important for discriminating between enantiomers!

6.6 Shielding

When we consider these nuclei and the molecules in which they reside, we will create a way to make atomic nuclei and molecular chirality work together to allow the two chiralities to be distinguished.

Figure 6.4. The alanine amino acid showing the L- and D-enantiomers. The arrows indicate the direction of the electric dipole moment (Berman et al. 2000). In this figure, the electric dipole moment lies almost entirely in the plane of the paper.

As discussed above, nuclei might achieve a strong magnetization if they are placed in the right conditions. This all happens independently of whatever the electrons that form the molecular bonds and give the molecule its shape are doing. The molecular bonds essentially move independently of the nuclei. However, the molecular electrons do indeed affect the nuclear alignment in a magnetic field via an effect employed in NMR and which can be applied in the present context.

Those molecular electrons enshroud the nucleus. The moving electrons create currents and thus generate their own magnetic fields at the nucleus. There is another effect going on as well if the molecules are moving through a magnetic field; this is known as the Lorentz force. Charges moving in magnetic fields feel an effective electric force acting on them. As the electrons move through the molecule, and as they are affected by the external magnetic field, they then generate their own magnetic fields within the molecule that tend to counter the external field.

Thus, the magnetic field at the nucleus is not exactly equal to the external magnetic field. It is modified by the molecular electrons. This effect is known as shielding (Buckingham 2004). Because this effect depends on the molecular shape, it is used in modern NMR as a sort of fingerprint for various molecules. However, even though all ^{14}N nuclei have the same magnetic moments, the shielding depends on the molecular configuration.

The electrons are not distributed uniformly throughout the molecule (which is why molecules have electric dipole moments), nor do they all move the same way. For this reason, the shielding is not uniform throughout the molecule for a particular nucleus. For the nitrogen nucleus in an amino acid, the shielding from electrons close to it may be greater than the shielding from electrons farther away. Also, the shielding depends on the orientation of the molecule in the magnetic field. The magnetic field at the nucleus (which we will refer to as the local magnetic field) is a vector and the external magnetic field is also a vector, and they are related in many ways. Thus, to describe the shielding, we need a mathematical tool that relates every component of the external magnetic field to every component of the local magnetic field. Since there are three components to each vector, we need nine components to relate all possible combinations of local and external field components. These combinations are related by a tensor, which links the local and external field components. In this way, each component of $\Delta \boldsymbol{B}$, the shift in magnetic field at the nucleus, depends on each component of the external field $\boldsymbol{B}$. As a result, each element of the tensor relates the three components of the magnetic field shift to the three components of the external field (x, y, and z).

What happens to the shielding tensor if there is an electric field? Because the electrons are charges, they are affected by the external electric field as well as by the magnetic field. This means that they behave differently from how they did before. The shielding tensor is changed even more by the presence of an external electric field (Buckingham 2004; Buckingham & Fischer 2006). In this case, the shielding tensor is a rank-3 tensor, meaning that it is three-dimensional. We must relate every component of the shift in the local magnetic field to every component of the magnetic field and every component of the electric field. Since each vector has three components, there are $3 \times 3 \times 3 = 27$ different combinations. For example, one component of the shielding tensor might be the shift in the magnetic field in the x-direction from the y-component of the external magnetic field and the z-component of the external electric field. Because the effects are small, an excellent approximation of the total shielding tensor is to write it as a sum of the tensor components, which depend only on the magnetic field, and a perturbation tensor, which depends on both the electric and magnetic fields. That is, we write this rank-3 tensor as an adjustment of the rank-2 shielding tensor. This perturbation tensor is called the nuclear magnetic shielding polarizability or the shielding polarizability for short.

For a mixture of molecules with random orientations, we can treat the shielding tensor plus the perturbed part in a sort of average fashion. Because molecules have a presumed zero magnetic moment, we treat them as being oriented randomly if there is only an external magnetic field. We can assume an isotropic distribution of molecular orientations with respect to the external electric and magnetic fields. The mathematics here results in a shielding tensor that is averaged over all possible combinations of molecular orientation with respect to the external electric and magnetic fields.

However, an electric field changes this. External electric fields tend to orient the molecule with its (possibly large) electric dipole moment aligned with that field. Of course, the temperature of the environment will tend to undo this effect, but unless the temperature is high, overall there will be a net alignment of the electric dipole moments in the direction of the external electric field. Because the molecule is aligned a certain way with respect to the external field, the shielding polarizability is no longer isotropic. Elements of the shielding polarizability in the direction of the external electric field, or the electric dipole moment since this is also aligned with the field, must be taken into account. Though this effect is temperature dependent, it can still be 100–1000 times larger than if the alignment were ignored. We saw from Figure 6.4 that the electric dipole moments are dependent on the molecular shape, which means that they are chirality dependent. Thus, the effect from the orientation of the molecules in the electric field on the shielding tensor depends on the molecular chirality. Because the shielding tensor is asymmetric, the orientation, hence chirality, of the molecule in the external fields matters.

Therefore, the external electric field changes the magnetic field at the nucleus. This means that the alignment of the nuclear spins with the external magnetic field also changes. Because the magnetic field is now coupled to the electric field, the nuclear spin is coupled to the molecular electric dipole moment. The net result is that the alignment of ^{14}N nuclei with respect to the external magnetic field in a molecule depends on the chirality of the molecule.

To summarize, magnetic fields orient the nuclei. Electric fields orient the molecule. When these fields exist simultaneously, the nuclei and molecules are reoriented with respect to each other. Further, the electric fields affect the magnetic fields at the nuclei in a way that is related to the molecular chirality. Thus, chiral molecules have a different nuclear orientation in this situation. *In external electric and magnetic fields, nuclei in molecules have magnetizations that depend on the chiral state of the molecule.*

This is a major piece of the puzzle in our search for the origins of meteoritic enantiomerism.

6.7 How Do We Get Magnetic and Electric Fields at the Same Time?

There are many places in the universe where the magnetic fields are quite high. Supernovae, neutron stars, black holes, and magnetars can have extraordinarily high surface fields (Bakirova & Folomeev 2016), and chunks of matter can get close to these objects. In fact, it may be possible to get too close; for high enough fields, the molecules may actually be ripped apart.

For our considerations, very large electric fields can be generated via Faraday's Law, which states that electric fields will be created if the magnetic flux changes. This effect may be present in the vicinity of a neutron star, or near a rotating neutron star, where the magnetic field lines change as well (Tsygan 1994; Timokhin 2007).

But there is a more likely way of generating electric fields for processing amino acids. This is the above-mentioned Lorentz force. This force is normal to the direction of the magnetic field and to the velocity of the particle. In the reference frame of the charge moving through a magnetic field, the force it feels would be akin to the force experienced from an electric field. This is known as motional EMF, and it is responsible for what is known as the translational Stark effect, described below.

In the reference frame of a meteoroid moving through a magnetic field, the molecules will feel both that magnetic field and an electric field due to the Lorentz force. This electric field is perpendicular to the magnetic field and to the velocity of the meteoroid. Because of this, the molecules have energies determined by their molecular electric dipole moment and the electric field in their reference frame (Rosenbluh et al. 1977; Panock et al. 1980; Zarnstorff et al. 1997).

There is another compelling possibility that is a combination of the above. Neutron stars can have huge magnetic fields. These fields take the form of a dipole field. If a meteoroid is moving through the field, molecules inside it experience an electric field though the translational Stark effect. However, because the magnetic field is nonuniform, the magnetic flux is also changing in the meteoroid's reference frame. The field's gradient about the neutron star can be extraordinary (Bakirova & Folomeev 2016), as much as 10^4 T km^{-1}. Because changing magnetic fields produce electric fields (Faraday's Law), a meteoroid moving through a magnetic field gradient can experience an electric field in its own frame of reference. The field gradients about neutron stars are amazingly large, so even slowly moving meteoroids can experience very large electric fields.

6.8 Nuclei Moving in Magnetic Fields, with Antineutrinos

In Chapter 5, we examined the mode by which ^{14}N nuclei might be selectively destroyed by neutrinos. This occurs via the following interaction:

$$\bar{\nu}_e + {}^{14}\mathrm{N} \rightarrow {}^{14}\mathrm{C} + e^+.$$

As discussed in Chapter 5, angular momentum conservation dictates that the probability for this reaction to occur depends on whether the spins of ^{14}N and the incident neutrino are aligned or antialigned. If it is the former case, then the probability for ^{14}N to be converted to ^{14}C is about an order of magnitude less than it is in the latter case (Boyd 2007; Wong 1998; Harwit 1981; de Shalit & Feshbach 1974; Morita 1973).

The antineutrino has a definite helicity, that is, its spin is always oriented in the same direction with respect to its momentum vector. This is shown in Figure 6.5. While neutrinos are oriented with their spin antiparallel to their velocity, antineutrinos have their spins oriented parallel to their velocity.

The nuclear spin, on the other hand, will be oriented according to the external magnetic field. The cross section varies smoothly as the angular alignment goes from perfectly anti-aligned to perfectly aligned. Thus, with the ^{14}N nuclei in a magnetic field and antineutrinos streaming from the source star, we can have a situation in

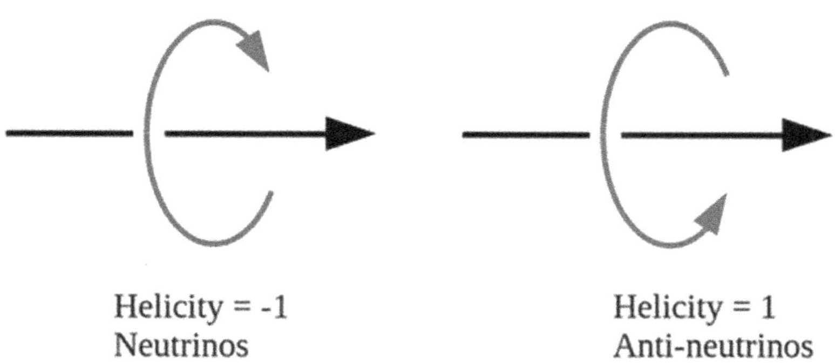

Figure 6.5. A representation of the neutrino helicity (left) and the antineutrino helicity (right). Neutrinos and antineutrinos have definite helicity. The horizontal arrows are the momenta of the particles, and the circles represent the spin.

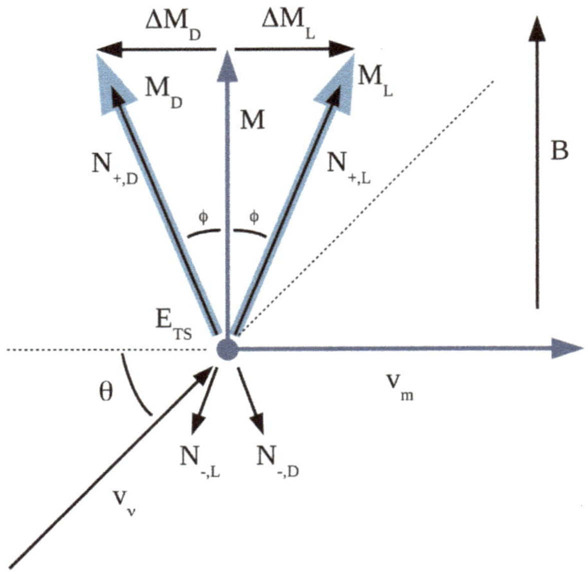

Figure 6.6. Vectors that are used in the SNAAP model with electric and magnetic fields. From Famiano et al. (2018).

which the nuclei have a specific alignment with respect to the antineutrino spin vectors and a cross section that depends on that alignment.

Figure 6.6 is a diagram of everything that is taking place in the SNAAP model. A meteoroid is moving with a velocity $\boldsymbol{v}_{\mathrm{m}}$ as shown in the figure. There is an external field $\boldsymbol{B}$, as indicated. The meteoroid does not have to be moving perpendicular to $\boldsymbol{B}$, although it is indicated that way in the figure. Because it is moving in a magnetic field, its velocity generates an electric field in the rest frame of the meteoroid. This is indicated by $\boldsymbol{E}_{\mathrm{TS}}$. In the figure, that vector is coming out of the page, represented by the dot in the center of the figure.

The magnetic field gives the nitrogen nuclei a net magnetization. Most nitrogen nuclei are going to have their magnetic moments aligned with the magnetic field, creating an average magnetization that points in the direction of $\boldsymbol{B}$. It is indicated by $\boldsymbol{M}$ in the figure. Note that not all nitrogen nuclei will have their spins in the direction of the magnetic field. Some will have their spins in the opposite direction. The proportion of those with spins parallel and antiparallel to the external field is determined by the external environmental temperature.

The neutrinos are streaming away from the neutron star, approaching each meteoroid with some velocity $\boldsymbol{v}_{\nu}$. The neutrino velocity is drawn at some average angle with respect to the meteoroid's velocity vector in such a way to indicate that the meteoroid is moving away from the star. This could represent a meteoroid in a hyperbolic trajectory that initially approached the star, shot past it, and then began to move away.

Consideration of what that electric field does to the nuclei and how it affects the magnetic field at the nucleus allows one to see how chirality and nuclear spin are related. Since chiral atoms have an asymmetric shielding tensor. The external electric fields change the magnetic field at each nucleus in a way that depends on the chirality. If the magnetic field at the nuclei is changed, so is the magnetization. This means that the vector $\boldsymbol{M}$ is adjusted for each chirality in opposite directions. In Figure 6.6, we denote the change in magnetization $\Delta\boldsymbol{M}_\mathrm{L}$ and $\Delta\boldsymbol{M}_\mathrm{D}$ for the left-handed and right-handed chiralities, respectively. The resultant magnetization for a chiral state is then $\boldsymbol{M}_\mathrm{L} = \boldsymbol{M} + \Delta\boldsymbol{M}_\mathrm{L}$ or $\boldsymbol{M}_\mathrm{D} = \boldsymbol{M} + \Delta\boldsymbol{M}_\mathrm{D}$.

Two details of this effect need to be considered. The first is that the change in magnetization is of opposite signs for the two chiral states, $\Delta\boldsymbol{M}_\mathrm{L} = -\Delta\boldsymbol{M}_\mathrm{D}$. The other is based on how the shielding tensor relates to the external magnetic field. A direct result of the shielding tensor calculation is that the vector $\Delta\boldsymbol{M}$ is perpendicular to both the external electric field and the unperturbed magnetization. The vectors $\Delta\boldsymbol{M}$ are shown this way in the figure. Incidentally, they are also pointing in the same direction for the amino acids alanine and isovaline, with $\Delta\boldsymbol{M}_\mathrm{L}$ parallel to the direction of velocity and $\Delta\boldsymbol{M}_\mathrm{D}$ antiparallel to it.

Thus, the resultant magnetization for each chiral state points in a slightly different direction. The vector $\boldsymbol{M}_\mathrm{L}$ points more in the direction of velocity, and the vector $\boldsymbol{M}_\mathrm{D}$ points slightly in the backward direction.

Since the magnetizations point in different directions, the averages of the different individual magnetic moments of nitrogen also point in different directions. The net result, though, will be affected by the thermal effects of our system. For a net magnetization, there will be a certain number of spins pointing in the direction of magnetization and a certain number pointing in the opposite direction. These are indicated by N_+ and N_-, respectively, in Figure 6.6. We have also subscripted these numbers with an "L" or "D" to indicate chirality. Thus, there are four possible combinations of chiral state and spin parallel or antiparallel to the external magnetic field. For a net magnetization, there will be fewer spins pointing in the direction opposite to the magnetization than in the direction of magnetization, as shown in the figure, where the arrow lengths indicate the total number of spins pointing in either direction.

Looking at this figure and recalling that the cross section for antineutrino capture on nuclei depends on the orientation of the nuclear spin with respect to the antineutrino velocity vector, we can deduce that the spin alignments with the largest destruction cross sections are those for the populations $N_{+\mathrm{D}}$ and $N_{-\mathrm{L}}$ as they are more antialigned with the antineutrino velocity vector. This depends on the angle θ that the antineutrino's velocity makes with the meteoroid velocity, but if we assume as before that the meteoroid is moving away from the neutron star, the selective destruction of $N_{+\mathrm{D}}$ over $N_{+\mathrm{L}}$ and of $N_{-\mathrm{L}}$ over $N_{-\mathrm{D}}$ will occur.

We also must take into account thermalization and magnetic fields. While we are destroying nuclei in particular spin states, thermalization will tend to randomize the spin states. At the same time, the magnetic field will attempt to maintain the equilibrium magnetization of the system. Also, since there are slightly more nuclei in the N_+ state, the rate of destruction of these nuclei will proceed faster. Since there are fewer nuclei in the N_- state, they will be destroyed at a lower relative rate. With all of these things going on at the same time, spin states will be simultaneously destroyed (but not at the same rate) and redistributed. But, how will all of these change the enantiomeric excess of the mixture?

6.9 Making All of This Work: Magnetochiral Effects with Neutrinos

Since, on average, the nuclear spins within the two chiralities will point in slightly different directions under the influence of the external magnetic and electric fields, the antineutrino interaction cross sections will be chirality dependent, while thermalization tends to drive things to random orientations, in contrast to the external magnetic field, which tends to drive nuclear spins to an orientation aligned with it. These effects on the nuclear orientation of a chiral state can be expressed by a set of coupled differential equations, one for the spin-up state and one for the spin-down state. There is one set of two equations for each chirality. We can treat each chirality independently since the model does not change the chirality of the molecule containing a spin state. It only destroys a particular spin state within a chirality. Before we get into the computations, it is worth mentioning a few places in the universe where we can simultaneously have magnetic fields, electric fields, neutrinos, and amino acids.

6.10 Potential Sites for the SNAAP Model

If the SNAAP model is to work, it needs two fundamental things: an external magnetic field and neutrinos. An electric field is also essential, but that will be produced in the reference frame of the meteoroid through its motion. However, two things work in opposition. First is the relaxation time of the ^{14}N nuclei, which moves spin-up nuclei into spin-down nuclei. Faster relaxation times mean that the nuclear spins are randomized more rapidly. The relaxation time acts as a throttle that governs how quickly things move along in the process of producing *ee*'s in an amino acid model. The other things that work in opposition to enantiomerism production are the variability in space and time of the neutrino flux and the magnetic field. Without neutrinos, the SNAAP model fails, as we will show in the next chapter. The magnitude of the flux depends on how far the meteoroid is from the source. Likewise, the magnetic field falls off as the cube of the distance from the source star.

6.10.1 Core-collapse Supernovae, Pros and Cons

It was previously mentioned that the prototypical site for the SNAAP model is a Type 2 supernova, or more specifically, a core-collapse supernova (Type 1b and 1c supernovae are also core-collapse supernovae). These produce 10^{53} ergs of neutrinos, which last, if those from SN 1987a are taken to be typical, on the order of a few seconds (Janka 2017; this is what we call the "fast-cooker" scenario).

Although core-collapse supernovae serve as the prototype for the SNAAP model, the red giant phase of their stellar evolution presents a huge problem, as noted earlier. The solution may be a massive star, a WR star, discussed in Chapter 3. The cores of these stars will also eventually become core-collapse supernovae. However, more massive stars are rarer than less massive ones, so there are not as many WR stars around as there are solar-sized ones. But there are enough of them to process a significant number of meteoroids having amino acids with enantiomeric excesses.

Core-collapse supernovae are not the only way for stars to produce strong magnetic fields and neutrinos. Once a neutron star has been produced and it has expelled its initial burst of neutrinos, it will still be hot, and so will continue to produce neutrinos for an expected duration of up to 10^5 years (Page et al. 2009). The rate will be far below that produced when the supernova explodes, around 10^{40} ergs s^{-1}, and the energies of the antineutrinos are uncertain. But, some situations may make that competitive with the supernova burst in processing nearby amino acids.

6.10.2 Neutron Stars

Core-collapse supernovae and neutron stars produce comparable magnetic fields of about 10^{12}–10^{15} G (or 10^8–10^{11} T) at the surface of the neutron star (or protoneutron star, in the case of a supernova). However, while the progenitor of a core-collapse supernova is about 1 au in radius during its red giant phase, the neutron star is a compact 10 km. Each system has its advantages and disadvantages. Even though the neutron flux for a cooling neutron star is small, meteoroids can get close to the source of both the field and the antineutrino flux, where their values will be higher. Also, because neutron stars cool for many years, meteoroids or planets can take their time to selectively destroy one enantiomer, particularly if they are in orbit around the star (what we refer to as the "slow-cooker" model).

Neutron stars had been thought to emit antineutrinos at thermal energies (Bahcall et al. 1974), which would give them insufficient energy to convert ^{14}N to ^{14}C. However, subsequent studies have shown that nuclear processes in cooling neutron stars will produce more energetic antineutrinos, many of which would be above the threshold to perform that conversion. Schatz et al. (2014) describe an Urca process (Nadyozhin 1995), or cycle, that occurs as the nuclei from the crust of the neutron star descends into the extremely neutron-rich matter of the star. As the nuclei progress toward the neutron-drip line, where additional neutron captures can no longer occur, the newly formed neutron-rich nuclei will decay by capturing an electron:

$$(Z, N) + e^- \rightarrow (Z - 1, N + 1) + \nu_e,$$

where (Z, N) is a nucleus with proton number Z and neutron number N. The next process that will occur is β-decay,

$$(Z - 1, N + 1) \rightarrow (Z, N) + e^- + \bar{\nu}_e,$$

which produces copious numbers of antineutrinos. Furthermore, since the nuclei involved are extremely neutron rich, the endpoint energies of the antineutrinos might be expected to be in the range of several MeV, quite energetic enough to convert ^{14}N to ^{14}C. However, the electron degeneracy in the star might force the emitted electrons to use most of the energy of the decay, leaving little for the antineutrinos.

However, other studies (Yakovlev et al. 2001) have suggested that processes not yet included in neutron star descriptions could produce higher energy antineutrinos.

Thus, although the use of a neutron star for SNAAP model processing appears to be uncertain, their antineutrino emission may yet turn out to be interesting.

Numerous neutron stars have been observed recoiling from their supernovae with a typical speed of 1000 km s^{-1}. That speed would allow them to pass through a vast amount of space during the time they continued to emit neutrinos, but it would not permit them to process very many meteoroids within the extended volume of a stellar system. If the neutron star captured a large meteoroid, it could process the material therein for a long time and might induce a relatively large *ee*. However, even if the meteoroid were large, it would not be expected to contribute much material to the interstellar medium. So, the recoiling neutron star is not expected to play much of a role in processing amino acids in passing meteoroids. The number density of processed meteoroids would likely not be sufficient to explain the observations from meteorites that make it to Earth.

However, consider the situation that would exist if a massive star and a neutron star were in a close binary system. Although these are not seen frequently in the Galaxy, they do exist. They produce X-ray bursters, which have been studied in the cases that have been found. In these systems, the two stars are well within the fraction of an astronomical unit that is the limit of the magnetic field when the supernova explodes and is presumably also close to the limit of the magnetic field that exists from the neutron star.

What usually happens in such systems is that the neutron star siphons off the outer layer (or perhaps two) of the massive star, creating an accretion disk around the neutron star. The disk, once formed, extends from just beyond the 10 km radius of the neutron star to distances that may extend to roughly 10^5 km (Popham & Sunyaev 2001) or more. The X-ray bursts are the result of occasional matter infalling onto the surface of the neutron star. The removal of the outer portion of the massive star prevents it from expanding to 1 au when it goes into its helium-burning phase. The disk would become thick, thus providing shielding for the material therein so that amino acids would be expected to form. The disks are also thought to be nurseries for creating meteoroids and even planets.

When the massive star explodes, all of the material in the disk, along with any planets that are within 1 au of the binary star system (and possibly a bit more, since the magnetic fields from both stars will orient the nuclei in the amino acids), will be processed by the antineutrinos from the supernova. However, the disk material might also have been processed for thousands of years by the antineutrinos being emitted from the neutron star (slower cooker scenario). Their intensity and energies would be lower than that produced by the supernova, but the material would be closer than the typical distance to the supernova, and that, together with the length of the processing time, might make the processing of the disk material by the neutron star before the supernova explodes the dominant *ee*-producing effect.

In addition, if the energies of the neutrinos and antineutrinos from the neutron star were sufficient, those energies might make them more selective than the supernova neutrinos in choosing left-handed amino acids over the right-handed ones. Few of the neutrinos, which would convert ^{14}N to ^{14}O, would exceed the threshold for that reaction, which is 5.14 MeV. The antineutrinos, which would convert ^{14}N to ^{14}C, would more easily exceed that reaction's threshold of 1.18 MeV.

Presumably, the binary system would initially have consisted of two massive stars, totaling, perhaps, 40 to 50 solar masses. As a result, when the second supernova exploded, there was almost that much mass in the surrounding vicinity, which would surely be greater in mass than a typical solar system. However, the material would not have expanded too much as, after being expelled by the two explosions, it would have been decelerated by the gravitational pull from the remaining two neutron stars. And when the two neutron stars spiraled into each other, more space debris, as well as more neutrinos, and many more heavy nuclei, would have been produced. There would be enough processed material in the vicinity to create its own stellar system and its own enantiomeric amino acids. If the binary system were created initially by the capture of a neutron star by the massive star, the total mass of the system would still be many times larger than that of the solar system.

Although binary systems, initially with two massive stars, are rare, only one such system was needed to produce life on Earth! And, either such a system or even a single WR star would be consistent with recent claims that our solar system was formed from a single WR star (Roberts et al. 2017). Indeed, this paper provides strong support for either of these scenarios.

Another interesting regime is that of neutron star mergers (or neutron star–black hole mergers). These events can have neutrino luminosities as high as those of Type II supernovae and the magnetic field strength in excess of that of a neutron star (Perego et al. 2014; Rosswog & Liebendorfer 2003; Dwarkadas et al. 2018). Moreover, because they are neutron stars, it is possible for meteoric debris to get close. In fact, it is conceivable that there is debris near these stars for two reasons. The first is that the supernovae that formed them will have expelled material in their vicinity. The second is the timescale for neutron star mergers to form—on the order of one billion years. This is plenty of time to make organic material and pack it into meteoroids before the neutron stars merged. These systems produce both the advantages of a core-collapse supernova and of a neutron star. We will have more to say about these systems in the next chapter.

References

Bahcall, J. N., Treiman, S. B. & Zee, A. 1974, PhLB, 52, 275

Bakirova, E. & Folomeev, V. 2016, GReGr, 48, 135

Berman, H. M., Westbrook, J., Feng, Z., et al. 2000, Nucleic Acids Res., 28, 235

Bloch, F. 1946, PhRv, 70, 460

Boyd, R. N. 2007, An Introduction to Nuclear Astrophysics (Chicago, IL: Univ. Chicago Press)

Buckingham, A. D. 2004, CPL, 398, 1

Buckingham, A. D. & Fischer, P. 2006, CP, 324, 111

de Shalit, A. & Feshbach, H. 1974, Theoretical Nuclear Physics I: Nuclear Structure (New York: Wiley)

Dwarkadas, V. V., Dauphas, N., Meyer, B., Boyajian, P. & Bojazi, M. 2008, LPI, 49, 1304

Famiano, M. A. & Boyd, R. N. 2016, in Handbook of Supernovae, ed. A. W. Alsabti & P. Murdin (Berlin: Springer), 2383

Famiano, M. A., Boyd, R. N., Kajino, T. & Onaka, T. 2018, AsBio, 18, 190
Famiano, M. A., Boyd, R. N., Kajino, T., et al. 2014, Symmetry, 6, 909
Harwit, M. 1981, MNRAS, 195, 481
Janka, H.-T. 2017, in Handbook of Supernovae, ed. A. W. Alsabti & P. Murdin (Berlin: Springer), 1575
Morita, M. 1973, Beta Decay and Muon Capture (Reading, MA: W. A. Benjamin)
Nadyozhin, D. K. 1995, SSRv, 74, 455
Page, D., Lattimer, J. M., Prakash, M. & Steiner, A. W. 2009, ApJ, 707, 1131
Panock, R., Rosenbluh, M., Lax, B. & Miller, T. A. 1980, PhRvA, 22, 1041
Perego, A., Rosswog, S., Cabezon, R. M., et al. 2014, MNRAS, 443, 3134
Popham, R. & Sunyaev, R. 2001, ApJ, 547, 355
Roberts, L. F., Lippuner, J., Duez, M. D., et al. 2017, MNRAS, 464, 3907
Rosenbluh, M., Miller, T. A., Larsen, D. M. & Lax, B. 1977, PhRvL, 39, 874
Rosswog, S. & Liebendorfer, M. 2003, MNRAS, 342, 673
Schatz, H., Gupta, S., Moller, P., et al. 2014, Natur, 505, 62
Timokhin, A. N. 2007, Ap&SS, 308, 345
Tsygan, A. I. 1994, A&AT, 4, 225
Wong, S. M. 1998, Introductory Nuclear Physics (New York: Wiley)
Yakovlev, D. G., Kaminker, A. D., Gnedin, O. Y. & Haensel, P. 2001, Phys. Rep., 354, 1
Zarnstorff, M. C., Levinton, F. M., Batha, S. H. & Synakowski, E. J. 1997, PhPl, 4, 1097

Richard N Boyd and Michael A Famiano

Chapter 7

How the SNAAP Model Selects Chirality

7.1 Computational Model

Previously, we have used the Buckingham effect (Buckingham & Fischer 2006) to calculate the effects of an external electric field on one amino acid. This resulting rank-3 tensor, known as the nuclear magnetic polarizability, was evaluated for L-alanine and D-alanine. We also calculated the effects of molecular orientation on the electric field and determined how this changed the effects of the external magnetic field and shielding tensor.

A molecular quantum chemistry code was used for these studies. Two popular codes today are Dalton (Aidas et al. 2014) and Gaussian (Frisch et al. 2016). There are also numerous other codes, each with their own advantages and disadvantages. The fundamental purpose of these codes is to solve numerically the Schrödinger equation for molecular electrons using a variety of techniques and under a variety of conditions. The solutions and molecular observables are usually calculated perturbatively. This is no small task, which is why quantum chemistry exists at the intersection of physics and chemistry, and has many expert practitioners.

7.1.1 Computational Methods

The details of quantum chemistry calculations are extensive, but we will mention a few of the fundamental pieces needed for the calculations, because we will be referring to these later on. These pieces are the computational method used to compute molecular properties and the basis sets used. We will first describe what we are calculating starting with very basic principles.

From a computational standpoint, a molecule is several electrons bound together in an array of attractive Coulomb potentials created by the atomic nuclei. From a chemical standpoint, the nuclei are treated very much like charged points (except in the cases when a few higher order corrections are needed). The electronic orbitals hold the nuclear array together. The shape of the molecule is determined by the energetic minimum of the electron wave functions.

doi:10.1088/978-0-7503-1993-5ch7

Within a molecule, the electrons are described by quantum-mechanical wave functions. Depending on the type of calculation, one might ascribe quantum numbers to these electronic wave functions. Because electrons are fermions, no two electrons can occupy the same wave function. This property is responsible for many of the chemical properties of a molecule, including shape, molecular binding energy, dipole moments, magnetic properties, optical properties, vibrational modes, and nearly every other chemical property.

The electronic wave functions do not just describe the distributions of the electrons in space, but they also describe their momenta. Once this information has been obtained, one can describe the energy states of a molecule as well as dynamic molecular properties such as current density within the molecule. By describing dynamic quantities such as electron momentum and angular momentum, it is possible to describe how the electron reacts to system perturbations such as magnetic and electric fields, both of which can change the molecular orbital characteristics. These changes in wave functions can affect how the molecular electrons interact with the nuclei.

When one assembles a quantum chemical calculation, there are three pieces that must be considered. The first is the computational method, that is, the mathematical tool used to find molecular orbital wave functions. Because solving the Schrödinger equation exactly for a system of electrons is impossible when one has more than two electrons, one necessarily has to resort to approximations.

One approximation method used is the Hartree–Fock approach (HF approach; Baym 1990). With it, the total electron molecular wave functions are described by a Slater determinant of orthonormal orbital functions. The determinant is fully antisymmetric, as is required for atomic electrons. The Hartree–Fock method starts with a trial wave function. Terms for the mean field potential from the other electrons are added. Molecular orbital coefficients are adjusted until the expectation value of the wave function energy is minimized. This process is iterative. Adjustments to the terms in the Slater determinant are followed by energy calculations until the energy reaches a minimum. A good description of the HF method can be found in most advanced quantum mechanics textbooks.

Quite often, the HF method suffers from a lack of accuracy because it does not include electron correlation effects. These effects can be included in the Møller–Plesset perturbation method (MP; Jensen et al. 1988), in which electron correlations are added by using the Rayleigh–Schrödinger (RS) perturbation theory (Lindgren 1974). Here, the unperturbed Hamiltonian is modified by a small external perturbation. In the MP theory, the initial wave functions are combinations of the eigenfunctions of the single-electron results. The perturbation is then added to include correlations. The perturbation to the Hamiltonian can then result in wave functions and energy computed to arbitrary order, and the method is often referred to as MPN, where N is the order of the perturbations used to obtain the solution (e.g., MP2 is a second-order perturbation, MP3 is a third-order perturbation, etc). Thus, an MP0 computation results in the eigenvalues and eigenfunctions of the single-electron Hamiltonian; it is the unperturbed solution. Generally, MP2, MP3, and MP4 methods are used, with the computational cost increasing with the order of the calculation. Significant accuracy can often be achieved even with MP2 calculations.

A third, computationally fast method is density functional theory (DFT). In this method, electron orbital wave functions are described by functionals (functions of another function). The functionals describe the spatial wave function of the electrons. One can imagine that the accuracy of the results depends heavily on the accuracy of the approximations made to obtain the functionals in this case. In DFT, the potentials are static (as usual in many-body calculations). Electronic stationary states are computed in an external field along with the electron–electron interaction energy (much like a usual Schrödinger equation). This many-body problem in DFT is treated like a single-body problem in which the electron density is solved for in a single wave function. In this case, the energy becomes a functional of the wave function, which can be minimized to find the electron density. Descriptions of DFT methods are given in Zhang & Musgrave (2007).

A common strategy is to first optimize the particle wave function using HF techniques, then take the solution to the HF optimization as input into an MP optimization. This saves time compared to solving the MP solution from scratch because the MP calculation then has a fairly accurate and reasonable starting solution from which to iterate. The solution of the MP optimization can be used as a starting point for the DFT to find the molecular properties of the system, using the DFT as just an algorithm to fit the functionals to the MP solution. This is the method we used for the results presented in this chapter, using an MP2 to optimize our molecular electron wave functions.

7.1.2 Basis Sets

The second piece necessary for computing electron wave functions is the collection of basis functions. These are used to describe individual orthonormal functions for specific computational methods. We fit the final electron wave functions to linear combinations of the basis functions. Think of the basis functions as a set of functions that we can combine to form the actual total electron wave function. These basis functions can be (though not always) steady-state solutions of the Schrödinger equation for single-electron orbitals. That is, they might be snapshots of the time-independent Schrödinger equation for specific quantized conditions. Each member of a basis set defines a unique solution to the Schrödinger equation. The actual total electron wave function is then a sum of one or more basis functions.

Because solving the many-body Schrödinger equation is subject to the handling of the computational method, the atoms involved, and the size of the molecule, there is no single set of basis functions that is universally used; there are actually several sets. The basis function sets are generally publicly available to the quantum chemistry community (Feller 1996; Schuchardt et al. 2007). The choice of basis set depends on the system size and composition, what properties one wishes to compute, and the computational cost. For example, for the amino acids, which consist of hydrogen, nitrogen, carbon, and maybe sulfur, we neither need nor want the Stuttgart RSC 1997 ECP basis set (Metz et al. 2000; Figgen et al. 2005, 2009; Peterson, et al. 2006,

2007), which has been written for the outer electrons of many elements heavier than oxygen. However, one may find that the 6-311G** basis set (which includes a limited number of orbitals for H, C, N, and O; Krishnan et al. 1980) is faster than, but not as accurate as, the aug-cc-pVDZ (Dunning 1989), which includes more basis functions than the 6-311G**. Basis sets can be applied to general molecular properties or optimized for specific purposes.

As with the computational method used, the basis set for our specific calculation can also be adapted to our desired calculation. For example, to examine the magnetic properties of molecules, one might optimize the electron orbital wave functions as above using an HF calculation with the 6-311G** basis set, which is very fast, and then re-optimize those results with an MP2 calculation using the aug-cc-pVDZ basis set. The resulting wave functions may be used to compute magnetic properties with a DFT calculation using a pcS-2 basis set, which has been developed specifically for magnetic properties (Jensen 2008, 2015).

With these pieces, molecular properties can be computed. Environmental properties are the third essential input into the calculations. A general computational sequence is diagrammed in Figure 7.1, along with other inputs that are required for this computation. While this is not the only sequence one may follow, this is the one we adopted to obtain the results presented in this chapter.

7.2 Shielding Tensor Calculations

To check the validity of our calculations, we compared our results to those of others. An initial comparison was done with the peroxide molecule (HOOH), which is chiral in certain configurations. As an example, we compared electrical and magnetic properties of the R–HOOH molecule (where the R indicates, as with amino acids, the handedness of the chiral HOOH) with a DFT calculation using the aug-cc-pVDZ basis set. Comparisons between the results of three different calculations are shown in Table 7.1. In this table, the isotropic shielding polarizability and the isotropic shielding tensors are also compared. The mathematical details of these are discussed in Appendix A. Such comparisons produce a useful test of the accuracy of our computational methods and choice of basis set.

With the successful comparison between our own calculations and prior calculations, we were able to proceed with confidence to the calculations of the shielding tensor for the amino acids.

7.2.1 Chirality-dependent Shielding in Amino Acids

A first amino acid calculation was performed for the L- and D-alanine (Famiano et al. 2018). As described above and in the appendix, the nuclear magnetic polarizability tensor was calculated. Recall that this is defined to be the effect on the shielding tensor in the presence of an electric field (Buckingham & Fischer 2006). As mentioned above, any of a number of readily available quantum chemistry codes can be used. Here, we used the Gaussian code (Frisch et al. 2016).

The meteoric environment is important but uncertain. In fact, there could be multiple environments in the interior of meteoroids. The amino acids could be in a

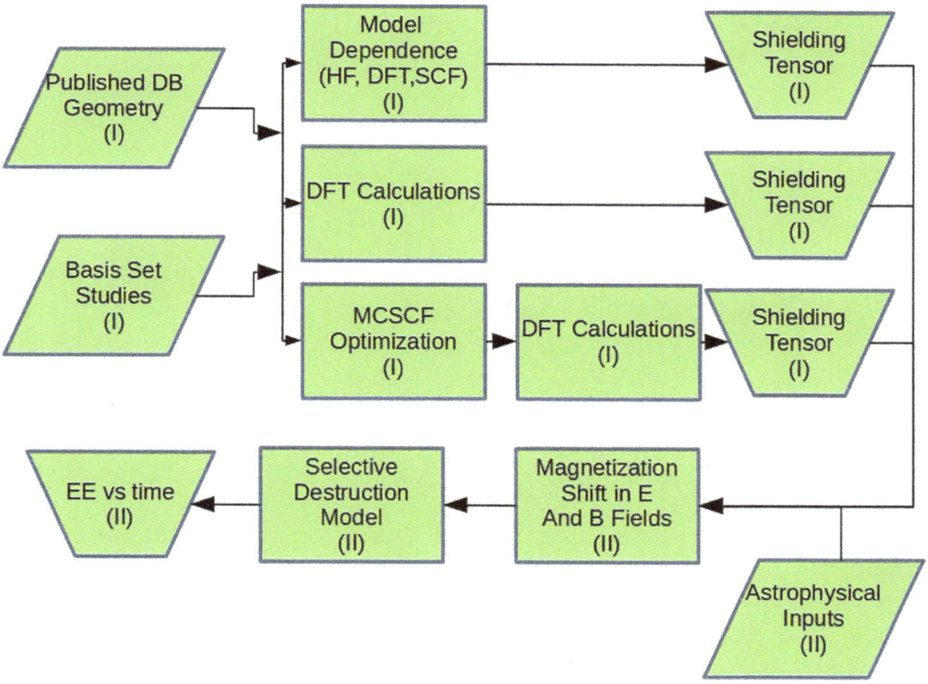

Figure 7.1. Schematic diagram of the sequence of calculations one might do in order to explore magnetochiral effects on molecules. The computational steps indicated by a "(I)" relate to calculations of molecular properties, while those indicated with a "(II)" indicate calculations in which the molecular properties are determined from different environmental conditions.

Table 7.1. Comparison of Isotropic Magnetic Shielding Polarizabilities and Nuclear Magnetic Shielding Constants for R–HOOH

	G1	G2	F1
$\overline{\sigma^{(1)}}^{(H)}$ (ppm a.u.)	−3.5 ± 0.43	−5.8 ± 0.19	−4.8
$\sigma^{(H)}$ (ppm)	26.3 ± 0.99	24.8 ± 0.60	26.68
$R_{O\text{–}O}$ (Å)	1.3866	1.452	1.3865
$R_{H\text{–}O}$ (Å)	0.9426	0.965	0.9426
∠OOH	103.059°	100.0	103.06°
∠HOOH	111.59°	112.0	111.46°

Note. Model G1 corresponds to Geometry 1 in Buckingham & Fischer (2006), and G2 corresponds to the MP2 result for Geometry 2 in the same reference. The model F1 corresponds to a DFT computation as described in the text (Famiano et al. 2018). Units are in ppm for $\sigma^{(H)}$ and ppm a.u. for $\overline{\sigma^{(1)}}^{(N)}$.

gaseous environment, in which they would be in ligand form, or they could be in an aqueous environment, in which their electron configuration would be zwitterionic. Further, the aqueous environment may have a nonneutral pH, in which case the enclosed amino acids may be cationic or anionic. In fact, some amino acids are naturally charged in aqueous environments. For this reason, shielding tensors were

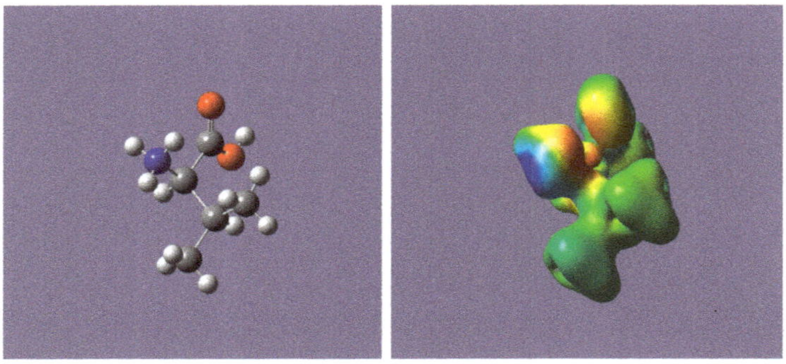

Figure 7.2. Left: the structure of the valine cation. Right: the structure of the valine current density and the shielding tensor for an external magnetic field in the x-direction. The surface is a contour of constant current density, and the colors are the strength of the σ_{xy} component of the shielding tensor for the nitrogen nucleus ranging from –0.1 ppm (red) to 0.1 ppm (blue). The orientation of molecules on the left and right sides of the figure is the same.

computed for several geometries of the same amino acid, corresponding to different environments.

We also adopted different basis sets depending on the calculation. For the optimization stage of the computation, a second-order MP2 computation (Jensen et al. 1988) was done using the aug-cc-pVDZ basis set (Krishnan et al. 1980), a commonly used basis set for atoms smaller than krypton. After optimization of the molecular geometry, we were able to compute the magnetic properties. For this, we used a DFT calculation using a B3LYP hybrid functional (Becke 1993) to describe the shape of our orbitals. We also used a pcS-2 basis set (Jensen 2008, 2015), a type of basis set designed especially for this sort of calculation.

A pictorial description of the shielding tensor for the nitrogen nucleus in cationic valine is shown in Figure 7.2, in which the structure of valine is shown next to a surface of constant electron current density induced by an external magnetic field in the x-direction. In this case, the surface maps out the locations where the induced electron current density is the same. This surface is colored according to the magnitude of the shielding tensor component σ_{xy} in an external magnetic field pointing in the x-direction.

7.2.2 Effects from the Molecular Electric Dipole Moment

The molecular electric dipole moment can have a significant synergistic effect on the shielding tensor in an external electric and magnetic field. Although we have described this model previously (Famiano et al. 2018), we will spend some time looking at the basic mathematics of amino acid processing by weak interactions here, referencing the vector diagram in Figure 6.6.

Remember that the electric dipole moment, i.e., the distribution of charge in the molecule, depends on the molecular shape; therefore, the electric dipole moments for the two chiral states will be mirror images of each other. Furthermore, the molecular orientation as a result of the electric dipole moment in an electric field can affect the shielding tensor, hence, the magnetic field, at the nucleus. Because dipole moments pointing in the direction of the electric field have a lower energy than those pointing opposite to the electric field, molecular electronic orbitals will have a preferred orientation in the external electric field. Temperature effects will result in a population distribution (Buckingham 2014; Buckingham et al. 2015) of molecular electric dipole moments aligned parallel and antiparallel to the external electric field based on Maxwell–Boltzmann statistics. We also learned that when the molecule moves in an external magnetic field, an electric field perpendicular to the magnetic field is produced in the molecular rest frame, as shown in Figure 6.6. The net result is that the electric field has an orientation that can be expressed relative to the magnetic field, the molecular electric dipole moment has a preferred orientation relative to the electric field, and the nuclear magnetic dipole moment has a preferred orientation relative to the external magnetic field plus its modification by the electric field. Thus, the nuclear magnetic dipole moment is coupled to the molecular electric dipole moment via the shielding tensor and hence to the molecular chirality.

For a molecule with an electric dipole moment $\boldsymbol{\mu}_E$ and a nucleus with a nuclear magnetization M in an external electric field $\boldsymbol{E}$, the shift in nuclear magnetization $\Delta_T M$ is (Buckingham 2014; Buckingham et al. 2015)

$$\begin{aligned}\Delta_T M &= (6kT)^{-1}[(\sigma_{xy} - \sigma_{yx})\mu_{E,\,z} + (\sigma_{yz} - \sigma_{zy})\mu_{E,\,x} + (\sigma_{zx} - \sigma_{xz})\mu_{E,\,y}] \times (M \times E)\\ &= M \times E(6kT)^{-1}\eta_M,\end{aligned}$$

where the σ_{ij} are the shielding tensor components and $\mu_{E,j}$ the molecular electric dipole moment components. From the above equation, it can be seen that there are two major terms multiplied by each other. One depends on the external environment and the other on the molecular geometry, consisting of the shielding tensor and the electric dipole moment. We refer to this latter term η_M as the molecular geometry factor:

$$\eta_M \equiv (\sigma_{xy} - \sigma_{yx})\mu_{E,\,z} + (\sigma_{yz} - \sigma_{zy})\mu_{E,\,x} + (\sigma_{zx} - \sigma_{xz})\mu_{E,\,y}.$$

For the sake of simplicity, we assume that the unperturbed magnetization M is perpendicular to the electric field $\boldsymbol{E}$. We can then write the relative shift in the magnitude of the nuclear magnetization ΔM as a product of an external environment term and the geometry factor:

$$\Delta M / M = E(6kT)^{-1}\eta_M.$$

It should be emphasized above that we have ignored the directions of the vectors M and ΔM in the above equation. Reminding ourselves of Figure 6.6 and the definition of the vector product, we see that the shift in magnetization is perpendicular to the unperturbed magnetization and parallel to the meteoroid velocity in the external magnetic field. The magnitude of this shift is proportional to η_M. If $\eta_M > 0$, then the magnetization shift is parallel to the direction of the meteoroid velocity, and if $\eta_M < 0$, ΔM is antiparallel to the meteoroid velocity.

From this, we can estimate the angle of the net magnetization $M + \Delta M$ with respect to the external magnetic field B. This is chirality dependent, since η_M is chirality dependent. The angle φ that the net magnetization makes with respect to the external magnetic field is then

$$\tan^{-1}\varphi = \Delta M/M = E(6kT)^{-1}\eta_M$$
$$\rightarrow \varphi \approx E(6kT)^{-1}\eta_M$$

since φ is generally small. Thus, the deflection angle of the magnetization is directly proportional to η_M.

The above is a result of the thermally averaged electrical polarization of a molecule oriented in a static external electric field. In this case, the product of the anisotropic components of the shielding tensor and the electric dipole moment vector results in a nonzero contribution to the change in magnetization. This nonisotropic effect can exceed the isotropic effects of the shielding tensor by several orders of magnitude (Buckingham & Fischer 2006).

If the meteoroid is not moving with respect to the magnetic field or if there is no external electric field, the bulk magnetization is M. For a moving meteoroid creating an electric field in the amino acid reference frame (or for some other electric field), an additional transverse magnetization component, $\Delta M_{\mathrm{L,D}}$, is created where the subscript indicates the L- or D-chirality. The chirality-dependent magnetization creates a shift in the nuclear spin vectors with populations $N_{\pm,\mathrm{L/D}}$ (see Figure 6.6). The magnetization shift then results in an angle between the average nitrogen spin and the antineutrino spin, which is chirality dependent. This produces the different reaction rates between antineutrinos and nitrogen nuclei, because nitrogen in each chiral state has a different spin alignment.

The reaction rate ratio for antineutrino spins parallel and antiparallel to the ^{14}N spin component is estimated (Famiano et al. 2018) to be

$$\begin{aligned} R_p/R_a &= (1 - \cos\Theta_p)/(1 - \cos\Theta_a) \\ &= (1 - \sin\varphi)/(1 + \sin\varphi) \approx (1-\varphi)/(1+\varphi) \\ &\rightarrow 1 - 2E(6kT)^{-1}\eta_M. \end{aligned}$$

Examining the case shown in Figure 6.6, we can now justify our previous statement that a positive η_M will result in a positive *ee* as the destruction rates for L-amino acids are smaller than for the D-amino acids. (For $90° < \varphi < 180°$, a negative value of η_M will result in an excess of L-amino acids.)

With the above destruction rate ratios, the time-dependent spin-state populations for each chirality, $N_{\pm,\mathrm{D/L}}$, and the enantiomeric excess of a particular state can now be computed,

$$ee = \frac{(N_{+,\mathrm{L}} + N_{-,\mathrm{L}}) - (N_{+,\mathrm{D}} + N_{-,\mathrm{D}})}{(N_{+,\mathrm{L}} + N_{-,\mathrm{L}} + N_{+,\mathrm{D}} + N_{-,\mathrm{D}})}$$

As a result, the enantiomeric excess is then coupled to the molecular orientation. Because the molecular orientation is controlled by the external electric field, the shielding tensor is affected by the molecular orientation in the magnetic field, and the nuclear orientation is affected by the resulting local magnetic field (Famiano et al. 2018) selectively destroying one enantiomer over the other.

7.3 Magnetochiral Effects with Neutrinos

Since different chiral states can have different magnetization vectors in electric and magnetic fields, on average the individual nuclear spins will point in directions that are based on the chiral state of the molecule. Because of this, the antineutrino capture cross sections are different for each chiral state. Of course, there are also thermal effects, which tend to drive things to random orientations, as well as the external magnetic field, which tends to drive things to an orientation aligned with it.

The nuclei in spin-up and spin-down orientations are affected by these environmental influences. The nuclear orientation of a chiral state can then be expressed by a differential equation, in fact, a set of two coupled differential equations. We have one equation for the spin-up state and one for the spin-down state. Recalling that we have one set of two equations for each chirality, we can treat each chirality independently since the antineutrino reactions do not change the chirality of the molecule containing a spin state. It only destroys a particular spin state within a chirality.

The math is detailed in the appendices, but the differential equations are shown schematically in Figure 7.3.

Each box in Figure 7.3 represents the number of nuclei aligned parallel or antiparallel to the external field. If there were no thermal effects, all of the nuclei would be preferentially aligned with the field in the N_+ state. When we add thermal effects, some of the nuclei aligned with the field will be placed in the antialigned state N_- against the field. The arrows for destruction by antineutrino capture are as indicated, and the relative lengths provide schematic insight as to which state has a higher destruction rate. Two things contribute to the destruction of a particular state: the alignment of the spin with respect to the neutrino momentum and the number of nuclei in that state. Using Figure 6.6, we determine that the destruction rates can be related as follows: $\lambda_{+,\mathrm{D}} > \lambda_{+,\mathrm{L}}$ and $\lambda_{-,\mathrm{L}} > \lambda_{-,\mathrm{D}}$. Also, because there are

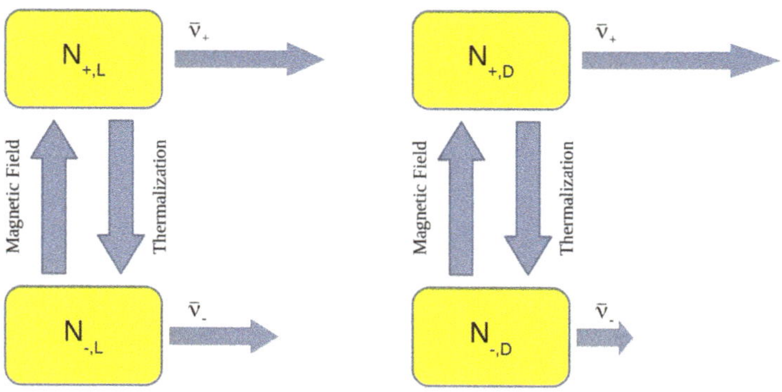

Figure 7.3. Schematic diagram of the differential equations governing nitrogen destruction in the SNAAP model. Both diagrams have similar terms that create and destroy a spin state, but the magnitudes of the terms differ. On the left is the diagram for the L-enantiomer, and the equivalent diagram for the D-enantiomer is shown on the right side. The rates of destruction differ for each chiral state and spin.

slightly more nuclei in the spin-up state, we know that $\lambda_{+,\mathrm{L}} > \lambda_{-,\mathrm{L}}$. Thus, we can rank the rate of destruction of nuclei in specific states as $\lambda_{+,\mathrm{D}} > \lambda_{+,\mathrm{L}} > \lambda_{-,\mathrm{L}} > \lambda_{-,\mathrm{D}}$.

Without neutrinos, the total number of nuclei is conserved. However, neutrinos destroy nuclei, so the total number of nuclei $N_+ + N_-$ decreases with time. As mentioned above, the rate at which N_+ decreases in time is not equal to the corresponding rate for N_-. Since these rates depend on the relative alignment of the magnetizations with the neutrino velocity vector, they depend on the molecular chirality. In summary, the rates different for the N_+ and N_- states, but the rate at which N_+ is destroyed for L-enantiomers is different from that for D-enantiomers.

The solutions to these sets of coupled linear differential equations enable the computation of N_+ and N_- for the L- and D-enantiomers as a function of time. The enantiomeric excess *ee* as a function of time is determined by subtracting the number of D-enantiomers from the number of L-enantiomers and dividing by the total number of molecules in the mixture. The resulting *ee,* of course, varies between −1 for a population of purely D-enantiomers to +1 for a population of purely L-enantiomers.

The solution for the number of chiral states as a function of time depends on the external magnetic field, the external electric field (which depends on the meteoroid velocity), the temperature, the shielding tensor, the angle between the antineutrino's velocity and the meteoroid's velocity, the antineutrino flux, the nuclear spin relaxation time, the angle between the meteoroid's velocity and the magnetic field vector, the molecular electric dipole moment, and the initial populations of the spin–chirality states. These are quite a few components to consider. Some of them we can measure, and some of them are stochastic (which is why the Monte Carlo code described in the previous chapter was written). Some depend on the environmental conditions. The relaxation time, for example, depends on the substrate or solution that contains the molecules (Nelson 2003). However, we have a reasonable idea of what it should be for each environment, or at least what reasonable values are (Troganis et al. 2003).

As an example, we will examine meteoroids in the vicinity of an antineutrino source (possibly a neutron star) carrying alanine and isovaline at 1% the speed of light in a 10 T field with an internal temperature of 10 K. The relaxation time for ^{14}N can be quite short, about 10 ms in aqueous solutions in the laboratory (Nelson 2003; Troganis et al. 2003). This means that it equilibrates thermally pretty rapidly after the antineutrinos destroy a certain spin state. This may also depend on several other factors. The relaxation time may well be longer in a frozen meteoroid.

For our initial calculations, we are particularly interested in the rate at which nitrogen is destroyed relative to the rate at which the spins relax. For this reason, we will define a relative factor, *f*, relating the antineutrino interaction rates, R_{p} (for spins parallel to the nitrogen spin), and the relaxation time, T_1. We will refer to *f* as the reaction-relaxation rate ratio,

$$R_p = f/2T_1.$$

The variability in the antineutrino capture rate depends on the distance of the meteoroid from the antineutrino source as well as the exact stage in the antineutrino

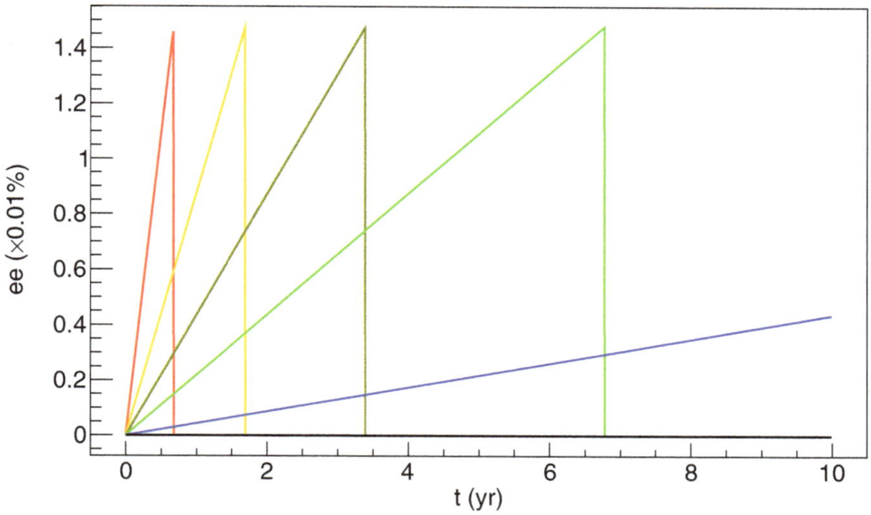

Figure 7.4. The enantiomeric excess of alanine as a function of time for a meteoroid in an external magnetic field for several long-term processing scenarios: slow, long-term $\boldsymbol{B} = 800$ T, $v = 0.015c$, and $f = 5 \times 10^{-8}$ (red), 2×10^{-8} (orange), 10^{-8} (yellow), 5×10^{-9} (green), 10^{-9} (blue), and 0 (black).

source's lifetime. We can envision a wide range of possibilities in which the rate changes dramatically (more on this later). Depending on the situation, the neutrino interaction rate can be several thousand times the relaxation rate or a tiny fraction of it. This ratio ultimately dictates how quickly the *ee* changes. This may be important as the time during which the antineutrinos are around can vary by a huge amount (on the order of a second for supernovae and possibly up to 100,000 thousand years for cooling neutron stars).

We can see some of the results of this model in Figure 7.4 for several assumptions of the ratio of antineutrino rates to relaxation rates for the amino acid alanine. The *ee* is seen to increase fairly linearly with time, then drops to zero after all of the molecules in the meteoroid have been destroyed. Depending on the antineutrino rate, this can occur on the order of a few seconds, which is about the time frame for antineutrino emission from a core-collapse supernova, to several years for a lower, but longer lasting, antineutrino flux.

For this situation, the *ee* is as high as about 10^{-4}. We also note the dramatic span in timescales. The time to maximum *ee* could be years or decades. This would require a meteoroid to orbit the source of the antineutrinos. For a higher-flux situation, the maximum *ee* may be reached after only a few seconds. In any case, the imbalance in chiral state populations is likely sufficient to seed autocatalysis, the process whereby even tiny *ee*'s can undergo further processing to produce 100% L-enantiomers (Soai et al. 1995; Soai & Sato 2002). In fact, laboratory experiments have resulted in autocatalyzing to 100% *ee*'s from about 10^{-7} (Soai, et al. 1995; Soai and Sato 2002; Soai, et al. 2014). We can see from Figure 7.4 that we are indeed in the ballpark.

Figure 7.4 also shows that the *ee*'s are the right sign! It is possible to conceive of a situation in which the correct sign of the *ee* is produced. We might imagine some orientations of $\boldsymbol{v}_{\mathrm{m}}$ and B that result in the opposite sign, but in general, for meteoroids moving away from the antineutrino source, the *ee* favors L-enantiomers.

It would make the SNAAP model more convincing if those *ee*'s were larger. After all, if the model can rely less on autocatalysis and more on the *ee* production mechanism, it would be more robust and autocatalysis would become less constrained. What sorts of magnetic fields and antineutrino rates would improve the model? Are there sites that could enhance the *ee*'s from the model?

Further, is alanine the best amino acid for producing large *ee*'s?

7.4 Results for Other Amino Acids

We can apply this calculation to amino acids other than alanine as well. Since the magnitude of this effect depends on the geometry factor η_M, we have compared η_M for several amino acids in both ligand and zwitterionic forms. The ligand geometries were taken from the Protein Data Bank (PC 2016), while the zwitterionic geometries came from the Cambridge Structural Database (Groom et al. 2016). The resultant values of η_M for several amino acids obtained from our optimizations are compared to literature values in Table 7.2 (units are in atomic units, in this case ppm a.u. or parts per million a.u.). Ligand and zwitterionic forms are shown along with some calculations for charged amino acids. In many cases, the zwitterionic forms have a much larger effect as a result of the larger electric dipoles in those forms.

In addition to the α-amino acids, we have also included in Table 7.2 the special cases of isovaline and norvaline. These are important because they are found in meteorites. In fact, isovaline is not found naturally on Earth, but it is found in meteorites with a positive enantiomeric excess.

This table also shows results for a few cases of amino acids in their zwitterionic forms as well as some in their cationic and anionic forms. The zwitterionic form is important as it results in a redistribution of charge in the amino acid, though the net charge remains zero. For example look at two different geometries for alanine in Figure 7.5. Because the charge is redistributed in the amino acid, a large electric dipole moment can result. We see from the equation describing the relative shift in nuclear magnetization that the dependence on the electric dipole moment, which can enhance the Buckingham effect, is increased. The zwitterionic form is the one that generally results in aqueous solutions. For nearly all amino acids in zwitterionic form, the value of η_M is positive, which would indicate that this natural form works well with this model.

There are a few notable cases, however. Both arginine and histidine have negative values of η_M in zwitterionic form, even for their optimized geometries. However, we should note that in their natural aqueous forms, these molecules have positively charged side chains. That is, they have an extra proton in their side chains. These positively charged cations are also calculated. In their natural cationic forms, arginine and histidine have positive values of η_M.

Table 7.2. Molecular Geometry Parameter η_M for Amino Acids Including Those Found in Meteorites

Amino acid	Ligand	Zwitterion	Optimized	Reference
Alanine	–3.87	31.79	39.39, **51.60**	LALNIN59, XIYSAA
Arginine	7.79	–44.11, **18.57, 47.18**	–160.41	TAQBIY, ARGHCL11
Asparagine	21.88	0.05, 38.33	–36.39, 37.53	VIKKEG, ASPARM09
Aspartic acid	11.22	11.83, 15.14	23.66	LASPRT04, NAGLYB10
Cysteine	10.24	8.92	10.60	LCYSTN36
Glutamic acid	0.94	28.48, 153.95	28.58	LGLUAC03, CAGLCL10
Glutamine	–4.76	16.43	9.22	GLUTAM02
Histidine	–10.55	–44.58, **23.26**	–31.20	LHISTD13, HISTCM12
Isoleucine	5.67	–5.24, 28.00	17.58	LISLEU02
Isovaline	–0.63	–1.92	–16.67, **119.94**	KIMKUO
Leucine	1.95	34.78	30.91	LEUCIN04
Lysine	0.53	0.09, 42.92, **–14.82**	19.78	CUFFUG, LYSCLH11
Methionine	–1.52	–0.34, 19.09	26.73	LMETON14
Norvaline	5.49	26.24	33.26, **10.50**	USOHUH04, VUKQID
Phenylalanine	12.10	19.3	21.15	QQQAUJ07
Proline	–3.68	17.5	47.25	PROLIN01
Serine	6.84	11.83	13.53	LSERIN28
Threonine	–6.43	12.20	–6.24	LTHREO04
Tryptophan	18.02	6.51	1.59	VIXQOK01
Tyrosine	19.72	28.37	–8.90	LTYROS11
Valine	1.01	4.44, 34.52	19.94, **8.47**	LVALIN05, VALEHC11

Note. Also included are geometry references from the CSD database (Pizzarello & Groy 2011) and the citation for amino acids used in this work. Also shown are the geometry parameters for several amino acids using the PDB geometries for the ligand form and the indicated reference from the CSD for the zwitterionic form. Values are in ppm a.u. Cations are in bold, and anions are underlined. Multiple values denote two indicated geometries in the same reference (e.g., two possible bond angles in the carboxyl group from the specified geometry).

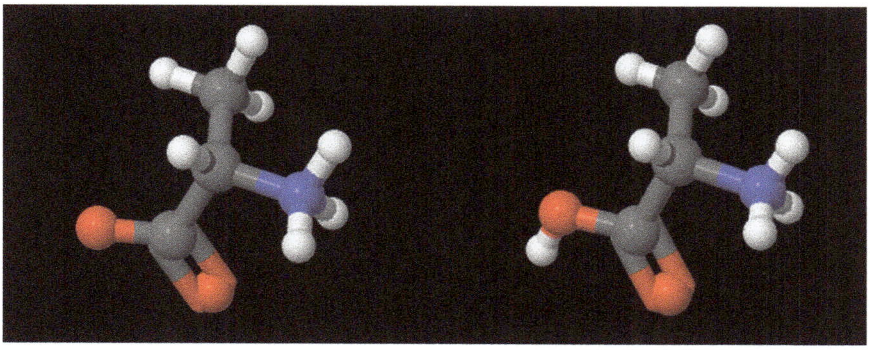

Figure 7.5. Different forms of the alanine amino acid. Left: zwitterionic form with the H from the carboxyl group moved to the nitrile group. Right: cationic form, with an extra H on the nitrile group. H is represented by white balls, C by gray, N by blue, and O by red.

Glutamic acid and aspartic acid are also interesting as their side chains are generally negatively charged in aqueous solutions. These anions have an extra electron in their orbital structure. In this case, the sign of η_M remains positive, though it is greatly enhanced for glutamic acid.

Isovaline, for all its importance (Cronin and Pizzarello 1997), remains mysterious. The value of η_M is negative for all three neutral cases. This situation is particularly interesting because the literature for the various forms of isovaline is sparse at best, with only a few special values reported for isovaline monohydrate (Butcher et al. 2013; Pizzarello and Groy 2011). It is interesting to note, however, that the cationic geometry of isovaline has a very large, positive η_M. This is compelling because it may suggest that in this state, the mechanism for a large *ee* of isovaline requires it to be in cationic form. This could occur in an environment with nonneutral pH, such as an aqueous hydrochloride environment. This is definitely worth exploring!

How well do these amino acids perform in the presence of external electric and magnetic fields? It is the Buckingham effect mentioned earlier that creates a chirality-dependent magnetization of the nitrogen nuclei in the amino acid. Using the results in Table 7.2 and a simulation of a meteoroid moving in a static magnetic field, we are able to evaluate this effect. For this simulation, we start with a racemic mixture (*ee* = 0) of isovaline. As with alanine, a nitrogen relaxation time of 10 ms (Troganis et al. 2003) and a neutrino momentum vector parallel to the meteoroid velocity vector (see Figure 6.6) were assumed. The temperature was taken to be 10 K. The reaction-relaxation rate ratio, *f*, was allowed to vary. It basically determines how fast the neutrinos are reacting with the atomic nuclei compared to the relaxation time.

The results of this calculation for cationic isovaline are shown in Figure 7.6 for four environments, which vary by their magnetic fields and range of *f*. These environments can be classified by the magnitude of the magnetic field (high or low) and the magnitude of *f* (high or low). These may even be representative of actual scenarios. For example, a meteoroid in hyperbolic orbit about a neutron star may have a magnetic field as high as 800 T, but a low *f* of $\sim 10^{-8}$. On the other hand, neutron star mergers or an accreting neutron star system may have a low field, $B \sim$ 30 T, but a high $f \sim 1$. Of course, the latter scenarios could also have much higher antineutrino fluxes, and possibly magnetic fields, in which cases their *ee*'s could increase to the level of percents.

The *ee* is plotted as a function of time in all four cases in Figure 7.6. Note that in the case of low *f*, the *ee*'s reach peaks of similar magnitude for the same magnetic field, but the time it takes to reach those peaks is much longer than in the case of high *f*, as would be expected. Of course, for $f = 0$, the mixture remains racemic as there are no neutrinos to selectively destroy amino acids. This indicates that the maximum *ee* in a given time depends on the magnitude of the magnetic and electric fields, while the rate at which the *ee* increases depends on the neutrino interaction rate.

Indeed, this is found to be the case. Figure 7.6 shows that this model can produce a maximum possible *ee* (for isovaline) in a certain amount of time depending on the neutrino flux and the magnetic field. How does this maximum vary with the

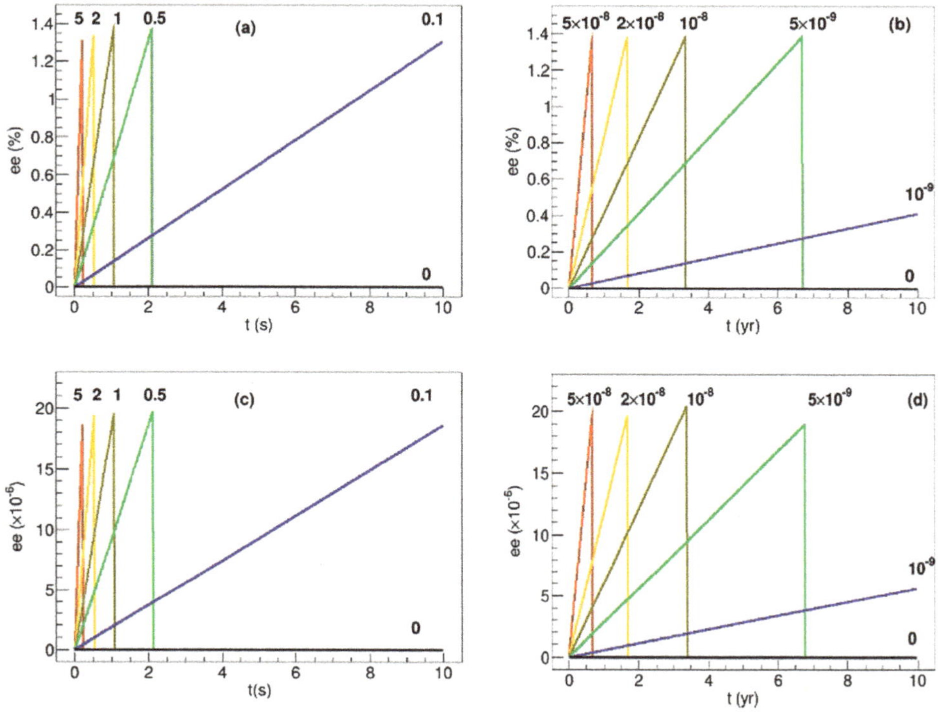

Figure 7.6. The isovaline cation *ee* as a function of time for a meteoroid penetrating the external magnetic field for scenarios described in the text, (a) $\boldsymbol{B} = 800$ T with high f, (b) $\boldsymbol{B} = 800$ T with low f, (c) $\boldsymbol{B} = 30$ T with high f, and (c) $\mathbf{B} = 30$ T with low f. Values of f are indicated beside the lines. In all cases, $v = 0.015c$.

parameters of the model? This is shown in Figure 7.7, in which the maximum *ee* is seen as a function of magnetic field and f, the ratio of the neutrino reaction timescale over twice the relaxation time. Recall that f is how fast the neutrinos are reacting with the atomic nuclei compared to the relaxation time.

In Figure 7.7, it is seen that the maximum *ee* depends strongly on the magnetic field, but not as directly on the value of the neutrino reaction rate. However, this figure also shows that the time it takes to reach the maximum does not depend strongly on the magnetic field, but does depend on the neutrino reaction rate. These are the two major parameters in determining optimal sites for the SNAAP model.

We have quantified our model with the geometry parameter η_M. In this case, the maximum *ee* produced increases with η_M. This can be used to get a rough estimate of the *ee* ratios for a particular site compared to those observed in meteorites. This is shown in Table 7.3, in which the ratio $\eta_M/\eta_{M,\mathrm{ALA}}$ (the ratio of η_M for a particular amino acid to that of alanine) is compared to the ratios of *ee* found in meteorites, ee/ee_{ALA}. In this table, ratios to both the Murchison and Murray meteorites (Pizzarello & Cronin 2000) are shown.

Here, negative values indicate that the amino acid being examined had a value of η_M of opposite sign to that of alanine, using the values in Table 7.2. Of course, it is difficult to draw any definitive conclusion from such a coarse comparison. The effect

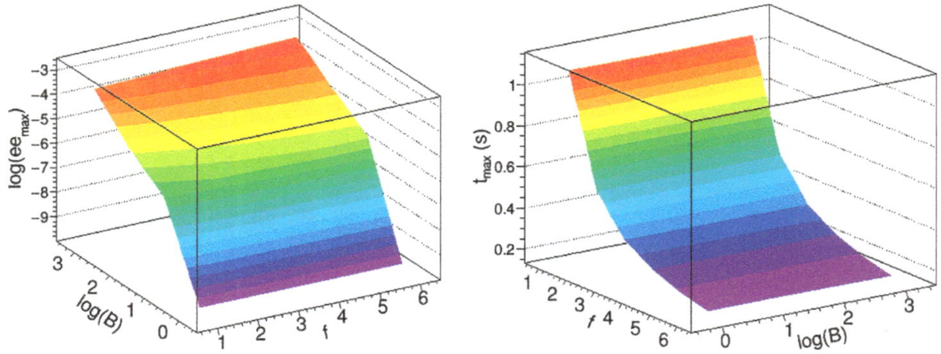

Figure 7.7. Left: the maximum enantiomeric excess, *ee*, for alanine as a function of the external magnetic field ***B*** and the value of *f*, assuming a meteoroid velocity of 0.01*c*. Right: the time it takes to reach the maximum *ee* as a function of *f* and the external magnetic field.

Table 7.3. Geometry Factor Ratios Compared to the Murchison and Murray Values (Pizzarello & Cronin 2000)

	Ligand	Zwitterion	Optimized Cation	Murchison	Murray
Isovaline	0.16	−0.06	2.20	7.00	3.5
Norvaline	−1.42	0.83	0.20	0.30	2.0
Valine	−0.26	0.14,1.09	0.16	1.83	−1.0

for the isovaline cation is seen to be significant compared to the other amino acids as well as in meteorites.

In fact, isovaline has been the poster child for finding amino acid *ee*'s in meteorites due to its significant values (typically on the order of a few percent). This supports the aforementioned hypothesis that the production of amino acids in meteorites is heavily dependent on the environment, which strongly influences the molecular geometry.

Table 7.3 also shows that even measurements are not necessarily consistent from one meteorite to the next. The ratios of *ee* vary dramatically and, in one case, the *ee* is in fact negative. Of course, the uncertainties in these measurements are still quite large (Pizzarello & Cronin 2000; Herd et al. 2011). It has also been found that aqueous alteration of meteoric material may dramatically influence the resultant *ee* (Glavin & Dworkin 2009). This is quite consistent with our calculations, which show that aqueous effects can be quite important.

Other effects may be worth exploring after amino acid processing takes place. Racemization of amino acids can occur, but it does so at different rates for different amino acids. Perhaps one reason why isovaline occurs with such a relatively high *ee* in meteorites is because it racemizes very slowly (Hudson et al. 2009). Also, subsequent autocatalysis may take place in meteoric environments. Finally, complex trajectories in sites of interest may change relative *ee*'s as well as the environmental conditions. Changes in pH and solution may result in a shift in the charge states of amino acids. However, our calculations have shown that amino acids can be chirally selected in environments with external magnetic and electric fields, as long as there are also electron antineutrinos.

There is one more issue we need to deal with. In this model, a ^{14}N nucleus is converted to a ^{14}C nucleus. The ^{14}C nucleus will β-decay back to a ^{14}N nucleus. Might that just return the amino acid to its original state?

When ^{14}C (with a half-life of 5730 years) decays by emitting an electron and an antineutrino, there are three particles involved in the product: the ^{14}N nucleus, the electron, and the antineutrino. Because ^{14}N is a more stable nucleus, there is 156 keV of leftover energy (the decay Q value). On the nuclear level, this is not a large amount of energy. However, on the molecular level, this is quite significant. Nuclear physics operates at energies of the order of MeV while atomic and molecular physics operate at energies of the order of eV.

The three particles involved in this decay share this energy. Each particle has a spectrum of possible energies after the decay. The ^{14}N nucleus can have as much as about 7 eV of energy as it recoils, from momentum and energy conservation. This is enough energy to completely dissociate the amine group from the amino acid. The energy of the amine group is in the neighborhood of 200–400 kJ mol^{-1}. For alanine, we can use 256 kJ mol^{-1} (Luo 2007). For one molecule, this is 2.66 eV. A ^{14}N nucleus recoiling with 7 eV of energy would be more than enough to dissociate the C–N bond. This effect (Snell & Pleasonton 1958; Oksyuk & Gerasimenko 1964) is likely to dissociate the molecule (Carlson 1960; Sassi et al. 2014).

However, the energy of ^{14}N depends on the kinematics, so that it may have very little recoil energy. In that case, the amino acid may return to its original state with its same chiral state. However, in this case, particularly for very long amino acid processing times, the molecule will simply be added back to the existing fuel for further processing. Eventually, a significant *ee* can result because of the differential destruction and recycling rate of the L and D forms. (For those that are not destroyed, a larger fraction will be added back to the fuel for subsequent processing, potentially resulting in a large buildup after a long time.)

Perhaps the most significant factor, however, is the recoil energy of ^{14}C when it is produced. The antineutrios that the ^{14}N nuclei capture would be of order 10 MeV. That would give the recoiling ^{14}C an energy of up to 10 eV, again depending on the kinematics. But even a fraction of this would be enough to destroy the molecule in which it was located.

There are also possible chemical and atomic results. When ^{14}N is converted to ^{14}C, the charge state of the molecule changes. The molecule will then either adjust its ionization state, or possibly its reactivity may change. Depending on the environment, the molecule may react or become unstable.

7.5 Sites for the SNAAP Model

We have defined f, the reaction-relaxation rate ratio, and the model's only free parameter r (although it assumes that the antineutrino energies are sufficient to perform the ^{14}N destruction), which varies with the environment in which the model occurs. At a distance of 1 au from a Type II SN, $f \sim 10^{-9}$. For a neutron star merger, f can be 10^{-4} or even much larger depending on meteoroid proximity. Cooling neutron stars may present another scenario. In this case, $f \sim 10^{-10}$ to 10^{-8}. The antineutrino reaction rate is expected to be much smaller for such sites. Despite this, the flux of antineutrinos could last for at least 10^5 years. The neutron star temperature decreases by about a factor of 3 during that time, although the neutrinos relevant to the SNAAP model may not be temperature related. Processing is possible as long as the magnetic field is present. If the meteoroid is in the vicinity of the neutron star for a few months or years, the amino acids may be processed, ultimately, to appreciable *ee*'s, provided the antineutrino energies are sufficient.

Any situation with a flux of energetic antineutrinos and high magnetic fields can result in amino acid processing. We can then use the reaction rate ratio for antineutrino spins parallel and antiparallel to the ^{14}N spin component as mentioned above to estimate the relative rates of amino acid destruction for each spin configuration. We also note from Figure 6.6 that the average nuclear spins are, for the most part, nearly perpendicular to the velocity vectors of the antineutrinos.

Table 7.4 shows parameters for several possible sites in which amino acids might be processed. The timescales given in this table are typical of those when the neutrino flux is present. Of course, in the case of a passing meteoroid, the astrodynamics would result in a changing field as it approached, which may enhance the *ee* even more.

As seen in Table 7.4, the SN II scenario is seen to not result in a large ee, simply because the meteoroid must be a large distance from the source of the antineutrinos in order to be outside the red giant star. Furthermore, with this scenario, there is the question of whether the nascent neutron star can evolve rapidly enough to achieve its full magnetic field before the neutrinos arrive at the meteoroid they are processing. However, the *ee* is so small that this is probably not a viable source of enantiomeric amino acids anyway.

There is at least one site that appears to be capable of processing meteoroids to close enough to the amount of enantiomeric amino acids that are found in the meteorites that make it to Earth to be interesting. This is a close binary system

Table 7.4. Properties of Sites Considered for the SNAAP Model

Site	Dist. (km)	Flux ($cm^{-2}s^{-1}$)	Rate ratio	B_0	B_r	Timescale	ee_{max}
SNeII	10^8	10^{30}	10^{-8}	10^{11}	10^{-12}	1	$<10^{-10}$
NSMerger	1000	10^{40}	0.1	10^{11}	500	1	0.1
WR Binary	1000	10^{34}	10^{-7}	10^{11}	500	10	10^{-5}

Note. These are Type II supernovae (SNeII), neutron star (NS) mergers, and WR Binary systems. The distance is a reasonable meteoroid distance from the object. The flux is the calculated antineutrino flux at the meteoroid. The rate ratio is the ratio of the antineutrino reaction rate to the relaxation rate. The magnetic field, in Tesla, is estimated at the surface, B_0, and at the meteoroid, B_r, using typical surface fields. The timescale is the approximate maximum time during which the antineutrino flux is present.

Figure 7.8. Artist's conception of a close binary stellar system, with one of the stars being a neutron star, and an accretion disk developing around it. Courtesy of European Space Agency, NASA, and Felix Mirabel (the French Atomic Energy Commission and the Institute for Astronomy and Space Physics/Conicet of Argentina).

consisting of a massive star and a neutron star (Boyd et al. 2018). The system presumably began as two massive stars, and the neutron star is the result of the completion of the stellar evolution of the more massive star. A conceptual picture of this is shown in Figure 7.8.

In a close binary, such as this is assumed to be, the outer one or two shells from the remaining massive star are attracted into an accretion disk, totaling several solar masses, around the neutron star, thereby converting the massive star into a WR star. The disk is sufficiently dense that dust grains, meteoroids, and even planets are thought to form there (D'Alessio et al. 2001). And surely, molecules such as amino acids would also form and would be processed by the antineutrinos from the massive star when it became a supernova and the magnetic field from the neutron star. The parameters used to calculate the *ee* of 10^{-5} shown in Table 7.4 are somewhat arbitrary, so that *ee* could be even higher. But admitting the possibility of autocatalysis does place this scenario in the status of possibly explaining the observed meteoritic *ee*'s.

The shock wave from the supernova would undoubtedly disrupt the disk after the processing had occurred, sending much of it, including a large amount of enantiomeric amino acids, into the interstellar medium (ISM).

The total mass of the two stars would probably be of order 40 solar masses. Such a system would certainly have enough mass to create our solar system; indeed, it has recently been asserted (Dwarkadas et al. 2017), based on anomalous abundances of radioactive elements, specifically ^{26}Al, that a WR star might have been the source of our system.

Calculating the *ee* anticipated from such a system is straightforward, assuming that the separation between the neutron star and the WR star is considerably larger than the size of the accretion disk. Ignoring the distortion of the disk's orbit that the WR star would impose (which would probably be large), we assumed that the two stars were separated by 10^6 km and the radius of the disk that contained amino acids was 10^3 km, putting it very close to the neutron star and to its intense magnetic field. The entire disk would still be roughly 10^5 km from the WR star when it became a supernova.

The amino acids in a meteoroid that was orbiting the neutron star in this system might also be processed by the magnetic field and the antineutrino flux from the neutron star as it cooled, provided that it can be shown by some future development that the antineutrino energies are sufficient to destroy the ^{14}N nuclei. Because the antineutrino flux from cooling neutron stars is so low, it would take a very long time to process the contained amino acids. In this case, however, time is the one thing that this situation has in abundance. The time span for neutron star neutrino and antineutrino cooling can be as long as 10^5 years (Yakovlev et al. 2001).

The neutron star merger case indicated in Table 7.4 produces the largest ee. In this situation, the initial phase might have been two massive stars. When one of them became a supernova and produced a neutron star, it would then proceed to suck the outer layer or two from the remaining massive star, converting it to a WR star and producing an accretion disk around the neutron star in which grains would form. Some of them would presumably also have amino acids formed on them. These would be expected to agglomerate into larger objects: meteoroids and even planets.

When the remaining star became a supernova, forming a second neutron star, the final stage would be set for them to merge, a situation described in Chapter 3. But as the two neutron stars closed in on each other, the disk that existed around the first one would become scrambled. The second neutron star would interact with large objects in the disk, dragging some with it as orbiting planets and propelling others into highly elliptical orbits. The net result would be a wide variety of objects, containing racemic amino acids, and spanning a range of distances from the neutrons stars as they closed in on each other.

When they finally underwent their explosion event, the antineutrinos they emitted would be mostly electron antineutrinos (Rosswog & Liebendorfer 2003), and they would be of about the same number and energies as those that would be emitted from a core-collapse supernova. Furthermore, they would apparently be emitted in a fraction of a second.

This event would produce an ee of order 10%, as in shown in Figure 7.9, about the level found in the meteorites that gave the largest *ee*'s.

The results of a calculation for a meteoroid at a range of radii in the neutron star merger model are shown in Figure 7.9. We see in this scenario that for a meteoroid moving in the vicinity of the newly created star to achieve *ee*'s in excess of 10%. These results are for alanine, and results for valine and isovaline may approach this same level of *ee*'s even sooner, depending on the system geometry.

Nonetheless, there are several caveats with both this result and that for the neutron-star–WR-star binary system. First, the magnetic fields in this region are

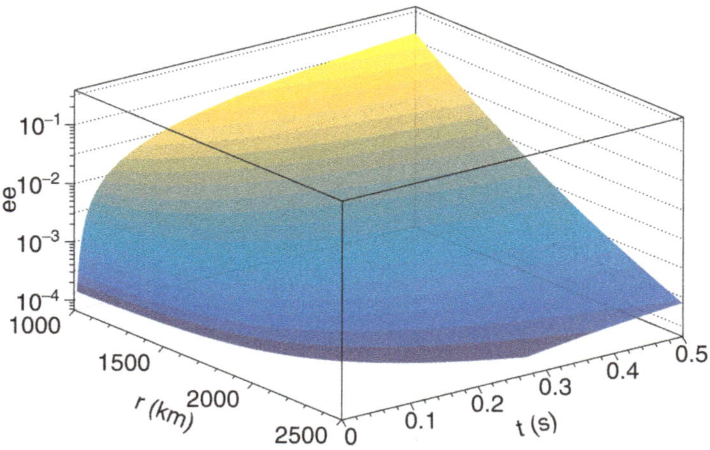

Figure 7.9. The *ee* as a function of time and meteoroid radius for the neutron star merger model for zwitterionic isovaline. The different colors represent the different *ee*'s that can be achieved. Note that large regions of parameter space exist that will produce *ee*'s of at least a few percent.

quite large. While the chosen surface field is realistic (though on the high end of the range of fields for neutron stars), even at 500–1000 km from the star the fields are intense. The same applies to the field assumed in the neutron star merger case (Rosswog & Liebendorfer 2003). This is worth noting for two reasons. First, the computational physics for computing magnetic properties of molecules generally employs first- or second-order perturbations to wave functions. For very high fields, this may not be sufficient.

Second, and closely related to the first point, it is not clear what happens to molecules at such high fields. We have a reasonable grasp of molecular behavior in fields of 10–100 T, but at several thousand Tesla or more, the electron orbitals may be deformed enough to reshape the atom or even destroy it, though studies on molecules in ultrahigh magnetic fields exist (Wunner et al. 1987) and progress is being made towards higher laboratory fields. Further, meteoroids moving in very large fields will experience large electric fields as well. While strong electric fields result in large *ee*'s, fields that are too strong may result in a dielectric breakdown of the material contained within the meteoroid.

However, Figure 7.9 does show that the conditions within our galaxy are quite capable of producing a large enantiomeric excess in amino acids. It appears that within the SNAAP model, autocatalysis may not even be necessary to produce meteoroids with the *ee's* found in those that have made it to Earth. Of course, amplification would be needed once the amino acids reached the surface of our planet, but that would certainly proceed given the high *ee* levels that this model seems capable of producing.

We should also note that planetary bodies in binary systems with accretion disks may have very complicated trajectories. This can provide a richness of possibilities within this environment. Accretion disk viscosity, field Lagrange points, and collisions can all change the meteoroid velocity. Also, the rotation of the neutron star can be responsible for the change in the magnetic field within the meteoroid. Thus, there are many possibilities within a single site, along with many combinations of field strength, trajectory, and electric field configurations. In some cases, these might even increase the *ee*'s predicted beyond those predicted from the current level of sophistication of the SNAAP model.

References

Aidas, K., Angeli, C., Bak, K. L., et al. 2014, WIREs Comput. Mol. Sci., 4, 269

Baym, G. 1990, Lectures on Quantum Mechanics (Menlo Park, CA: Addison-Wesley)

Becke, A. D. 1993, JChPh, 98, 5648

Boyd, R. N., Famiano, M. A., Onaka, T., Kajino, T. & Mo, Y. 2018, ApJ, 856, 26

Buckingham, A. D. 2014, JChPh, 140, 011103

Buckingham, A. D. & Fischer, P. 2006, CP, 324, 111

Buckingham, A. D., Lazzeretti, P. & Pelloni, S. 2015, MolPh, 113, 1780

Butcher, R. J., Brewer, G., Burton, A. S. & Dworkin, J. P. 2013, AcCrE, 69, o1829

Carlson, T. A. 1960, JChPh, 32, 1234

Cronin, J. R. & Pizzarello, S. 1997, Sci, 275, 951

D'Alessio, P., Calvet, N. & Hartmann, L. 2001, ApJ, 553, 321

Dunning, T. H. Jr. 1989, JChPh, 90, 1007

Dwarkadas, V. V., Dauphas, N., Meyer, B., Boyajian, P. & Bojazi, M. 2017, ApJ, 851, 147

Famiano, M. A., Boyd, R. N., Kajino, T. & Onaka, T. 2018, AsBio, 18, 190

Feller, D. 1996, JCoCh, 17, 1571

Figgen, D., Peterson, K. A., Dolg, M. & Stoll, H. 2009, JChPh, 130, 164108

Figgen, D., Rauhut, G., Dolg, M. & Stoll, H. 2005, CP, 311, 227

Frisch, M. J., Trucks, G. W., Schlegel, H. B., et al. 2016, Gaussian 16, Revision A.03, (Wallingford, CT: Gaussian Inc.)

Glavin, D. P. & Dworkin, J. P. 2009, PNAS, 106, 5487

Groom, C. R., Bruno, I. J., Lightfoot, M. P. & Ward, S. C. 2016, AcCrB, 72, 171

Herd, C. D. K., et al. 2011, Sci, 332, 1304

Hudson, R. L., Lewis, A. S., Moore, M. H., Dworkin, J. P. & Martin, M. P. 2009, Bioastronomy 2007: Molecules, Microbes, and Extraterrestrial Life (San Francisco, CA: Astronomical Society of the Pacific)

Jensen, F. 2008, J. Chem. Theory Comput., 4, 719

Jensen, F 2015, J. Chem. Theory Comput., 11, 132

Jensen, H., Aa, J., Jørgensen, P., Agren, H. & Olsen, J. 1988, JChPh, 88, 3834

Krishnan, R., Binkley, J. S., Seeger, R. & Pople, J. A. 1980, JChPh, 72, 650

Lindgren, I. 1974, PhLB, 7, 2441

Luo, Y.-R. 2007, Comprehensive Handbook of Chemical Bond Energies (Boca Raton, FL: CRC Press)

Metz, B., Schweizer, M., Stoll, H., Dolg, M. & Liu, W. 2000, Theor. Chem. Acc., 104, 22

Nelson, J. H. 2003, Nuclear Magnetic Resonance Spectroscopy (Upper Saddle River, NJ: Pearson Education)

Oksyuk, Y. D. & Gerasimenko, V. I. 1964, JETP, 46, 254

PC 2016, National Center for Biotechnology Information, Pubchem Compound Database; cid=5950, https://pubchem.ncbi.nlm.nih.gov/compound/5950 (accessed Nov. 5, 2016)

Peterson, K. A., Figgen, D., Dolg, M. & Stoll, H. 2007, JChPh, 126, 124101
Peterson, K. A., Shepler, B. C., Figgen, D. & Stoll, H. 2006, JPCA, 110, 13877
Pizzarello, S. & Cronin, J. R. 2000, GeCoA, 64, 329
Pizzarello, S. & Groy, T. L. 2011, GeCoA, 75, 645
Rosswog, S. & Liebendorfer, M. 2003, MNRAS, 342, 673
Sassi, M., Carter, D. J., Uberuaga, B. P., Stanek, C. R. & Marks, N. A. 2014, BBA, 1840, 526
Schuchardt, K. L., Didier, B. T., Elsethagen, T., et al. 2007, J. Chem. Inf. Model., 47, 1045
Snell, A. H. & Pleasonton, F. 1958, JPhCh, 62, 1377
Soai, K., Kawasaki, T. & Matsumoto, A. 2014, Chem. Rec., 14, 70
Soai, K. & Sato, I. 2002, Chirality, 14, 548
Soai, K., Shibata, T., Morioka, H. & Choji, K. 1995, Natur, 378, 767
Troganis, A. N., Tsanaktsidis, C. & Gerothanassis, I. P. 2003, JMagR, 164, 294
Wunner, G. & Ruder, H. 1987, PhyS, 36, 291
Yakovlev, D. G., Kaminker, A. D., Gnedin, O. Y. & Haensel, P. 2001, PhR, 354, 1
Zhang, G. & Musgrave, C. B. 2007, JPCA, 111, 1551

Chapter 8

Experimental Tests of the SNAAP Model

8.1 Possible Measurements for Direct Confirmation of the SNAAP Model

We described experiments that were performed to test the circularly polarized light (CPL) model of amino acid chirality selection, but we have not presented any data in support of the SNAAP model. The probabilities for antineutrinos or neutrinos from man-made sources to interact with matter are extremely small, so the acquisition of data resulting from neutrino-induced reactions requires pretty heroic experiments. And, of course, neutrino sources are really particle accelerators—and not little ones either. Neutrino–nucleus interaction probabilities, cross sections to physicists, have been measured for a number of nuclei, but no one has tried to infer how they might affect the net chiralities of amino acids. Although an experiment using a neutrino beam would be difficult, it might be possible. However, we believe that more straightforward experiments might be conducted to at least test some of the ideas that are basic to the SNAAP model.

8.1.1 Measurements with the Spallation Neutron Source

We will first describe possible experiments using neutrino beams. One obvious place to do an experiment, both because the neutrino beam would be quite intense there and because the energies of the neutrinos and antineutrinos, several tens of MeV, would be roughly comparable to those that would be emitted by a supernova, would be the Spallation Neutron Source (SNS) at Oak Ridge National Laboratory in Tennessee. This facility was built to produce a neutron beam to be used in studying neutron-induced processes, but it also will produce an intense neutrino beam. The primary driver for this facility is a medium-energy proton beam, which produces copious numbers of positively charged pions (see Chapter 5) through its interactions with target nuclei. These will decay to muons and a muon neutrino or antineutrino,

doi:10.1088/978-0-7503-1993-5ch8

which will then decay to a positron or electron and two more neutrinos. The positrons will annihilate with electrons to produce photons.

The resulting time distribution of the different neutrino flavors that will be produced by the SNS is shown in Figure 8.1. This is especially important in distinguishing between events initiated by the different flavors of neutrinos, or between a neutrino and its own antineutrino. The number of muon neutrinos is huge, but it is also prompt, so any reactions they might induce can be distinguished from their time distribution. Unfortunately, the electron antineutrino flux is seen to be extremely small, so they could not be used to perform a useful experiment.

So what would an experiment to test the SNAAP model look like? The most direct experiment would be one in which the electron antineutrinos or neutrinos would bombard a sample of a particular amino acid, which would be encased in as strong a magnetic field as could be generated with a laboratory magnet and an imposed electric field. The conversions of the nitrogen nuclei to carbon or oxygen can then be measured using any of a number of available techniques, such as NMR. It would also be important to reverse the magnetic field at some point to see if the reaction rate would be affected (the neutrino spin direction would remain the same). Likewise, the experiment can be performed on separate samples of L- and D-amino acids to measure the relative amounts before and after exposure. The crossed electric field can be created within the sample using a parallel plate array. Laboratory DC fields of several million V m^{-1} or more can be achieved. This would necessarily be a fairly compact setup to accommodate the bore of the external magnet. Such a setup can also be tuned to the field strength and direction relative to the external magnetic field and the neutrino momentum vectors. This would provide a way to simulate various conditions of the meteoroid within the vicinity of the neutron source.

Figure 8.1 shows the intensities of the various neutrino flavors expected from the SNS (Avignone & Efremenko 2003), as well as the time distribution. Although it should be straightforward to discriminate between events initiated by electron

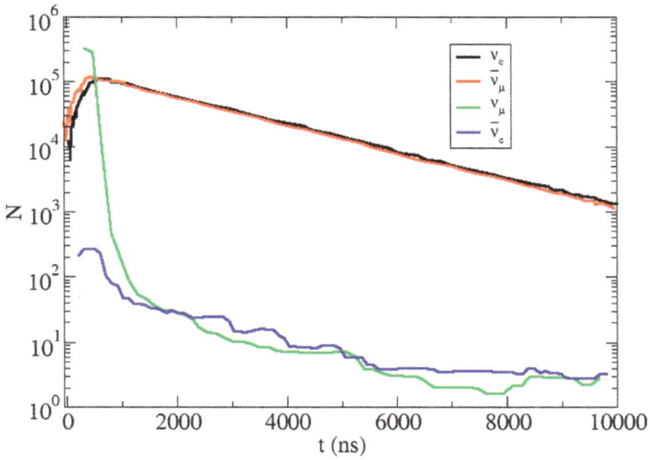

Figure 8.1. Arrival times and rates of the different neutrino flavors produced by the Spallation Neutron Source. ns = nanoseconds = billionths of a second. Data derived from Avignone & Efremenko (2003).

neutrinos and mu neutrinos, it is unfortunate that the intensity of electron antineutrinos is so small. This does not really present a problem for testing the SNAAP model, however. The electron neutrino energies are high enough that the relatively large negative threshold energy for them to convert ^{14}N to ^{14}O should not present as much of a problem as it does for supernova neutrinos or neutrinos from coalescing neutron stars. Furthermore, the reaction cross section for converting ^{14}N to ^{14}O should be essentially the same as that for converting ^{14}N to ^{14}C. Thus, this experiment, if feasible, should produce an excellent test of the SNAAP model viability.

This, however, is not to say that this experiment would be easy. Neutrino cross sections are so tiny that enormous care needs to be taken to ensure that backgrounds from cosmic rays do not produce spurious events. Perhaps the most insidious background would be the cosmic neutrinos that are produced in Earth's atmosphere. The solar neutrinos, mentioned in Chapter 3, would be too low in energy to cause much of a background. Nevertheless, all such sources would have to be considered before even attempting this experiment.

While this would be a superb test of the SNAAP model's ability to select one chirality, the experiment would undoubtedly require years to acquire enough data to show convincingly that the SNAAP model would select one chirality over the other and, furthermore, that it would be the correct chirality!

8.1.2 Measurements with Neutrinos from a Nuclear Reactor

Another possible source of neutrinos could be reactor neutrinos. Reactor antineutrino spectra cover a range to roughly 8 MeV with a peak at around 3 MeV. This is sufficient to convert N to C in amino acids. Another interesting aspect of the reactor antineutrinos is that, because they are produced by β-decay of neutron-rich nuclides, the antineutrino flux is the only neutrino flux. There are no electron neutrinos or other types of antineutrinos. Thus, the location of amino acid samples in or near the reactor would provide a sizable flux of electron antineutrinos.

Research reactors are capable of situating samples in the reactor core. However, these are generally small sample containers. Because an external magnetic and electric field must also be provided for this model to work, a fairly large device (e.g., a dipole magnet) must be set up. Even superconducting magnets are quite large, and producing a magnetic field in excess of 10 T would require a special magnet indeed. In addition, an apparatus to generate an electric field of one to several million V m^{-1} must be implemented. Generating such a field could be done in a small space, allowing the sample and field generator to be placed within the magnet bore. Even placing the magnet 10 m from the 3 GW reactor is estimated to generate a neutrino flux four orders of magnitude higher than the flux generated from the SNS (An 2017). This could be a commercial reactor, access to which may be limited. Research reactors would have an antineutrino flux one to two orders of magnitude lower than this but still higher than the SNS flux. Note that the energies of the antineutrinos from a reactor would be a bit lower, but of the same order as those anticipated from supernovae or coalescing neutron stars.

Perhaps the simplest way to conduct the test of the SNAAP model would not be to observe the change in chirality in real time. Instead, a general prescription for doing such an experiment might be the following:

- Put a sample of racemic amino acids in the magnetochiral device, i.e., a region with magnetic and electric fields, in the flux of neutrinos or antineutrinos from either a reactor or accelerator.
- Then, after an exposure, remove the sample and measure the *ee* via any of a number of possible methods, many of which are based on gas chromatography methods.

An alternative would be to use purely L- and D-amino acid samples and measure the amount of each before and after irradiation, providing a measure of the destruction.

For the postirradiation analysis of amino acids, one possible method is multidimensional gas chromatography (Myrgorodska et al. 2016). This method is highly sensitive, with detection limits of as low as 0.005 pmol (about 3 billion molecules). This method generally is limited by the amount of amino acid present as opposed to the ability to separate enantiomers. Typical measurement resolution could be a challenge as uncertainties must be well under 1%, though this might be the same order of magnitude as the expected *ee* in some cases. Contributions to measurement uncertainties also include the small sample size. By starting with a relatively large sample, the experimental production of amino acid *ee* could possibly be measurable.

While the results presented in Chapter 7 suggest that cationic isovaline may exhibit the largest effect in the SNAAP model, it would be worthwhile to examine several amino acids in a variety of environments. This would provide not only a test of the SNAAP model, but also a comparison of the relative magnitudes of the effects for several amino acids.

8.2 Nuclear Magnetic Resonance Measurements as Tests of the SNAAP Model

The SNAAP model opens a new door in the field of physical chemistry. Chirality differences in the nuclear magnetic polarizability from the Buckingham effect can be computed (Buckingham 2004; Buckingham & Fischer 2006). Recall that it is this polarizability that alters the magnetic field at the nucleus of the atom in a molecule, and this polarizability depends on the chirality of the amino acid. This effect is further enhanced by the alignment of a molecule in an external electric field. Because amino acids generally have strong dipole moments, their alignment in an external electric field is more enhanced than is typical.

It would be worthwhile to verify the magnitude of this effect in amino acids. While calculations are thought to be fairly accurate, the SNAAP model does subject its molecules to extremely high magnetic and electric fields, so experimental confirmation would be welcome. The mathematics behind this effect is described in Appendix A. Experimental measurements of this effect could possibly be done in an NMR experiment.

In such a measurement, a substance is placed in a large magnetic field. Nuclei with nonzero spin (usually hydrogen, but not always) will have a population that aligns with the field and a population that aligns antiparallel to the field. These are the low- and high-energy states of the nuclei. By applying an external oscillating field, nuclei in the low-energy state receive enough energy to be placed in the high-energy state. However, because the environment is at a nonzero temperature, the nuclear spins will equilibrate with a timescale equal to the relaxation time of the substance. This relaxation time depends on the material, the solution, the structure of any crystal, and several other factors.

For an external field $\boldsymbol{B_0}$ pointing in the z-direction, the energy difference between the two spin states of the nucleus ΔE is

$$\Delta E = 2m_s\hbar\gamma B_0,$$

where γ is the gyromagnetic ratio of the nucleus, that is, the ratio of the magnetic moment to the angular momentum, and m_s is the magnetic substate. For a spin-1/2 nucleus, like hydrogen, then

$$\Delta E = \hbar\gamma B_0.$$

By applying an external energy source, such as an RF field, to the sample, nuclei in the lower energy state can be placed in the upper spin state. If the external source is precisely the amount of energy needed to promote the nuclei in the lower spin state, then a resonant condition is met. The required frequency ν of the external RF field is

$$h\nu = \Delta E.$$

When the nuclei de-excite, they give off energy, which can be detected with an RF receiver. Because the nuclei de-excite with a characteristic time constant, then an RF signal, which decays in time, is detected. The applied external RF signal is pulsed, and then the RF receiver detects the signal decay. This is why nitrogen is not a good nucleus to look at in NMR. Although it is possible, the relaxation time of nitrogen is so small that it is difficult to detect quickly enough after the initial pulse. (Nitrogen also has a very broad chemical shift, meaning that it is difficult to pick out of an NMR spectrum.) The frequency at which a resonance occurs is the characteristic observable in an NMR experiment. Note that this describes a simplified view of Fourier Transform (or FT) NMR. Other ways of conducting NMR experiments are possible, but this description is sufficient for what follows.

Molecular structures can be identified in NMR by examining the frequency at which a resonance occurs. It would seem that identifying, for example, hydrogen in an NMR experiment to identify specific chemical structures is insufficient because all hydrogen nuclei (protons) have the same magnetic moment and gyromagnetic ratio. This is where molecular structure comes in. The molecular electrons create a shielding effect (described in the previous chapter) for the nuclei, which depends on the configuration of electrons in a molecule. This effect reduces the magnetic field at the nucleus, and thus the actual magnetic field. Hence, the NMR resonant frequency is shifted somewhat based on the structure of the molecule. This is referred to in the NMR world as the chemical shift, which acts as a fingerprint for identifying molecules. Other effects include the spin–spin coupling, by which spins of nuclei affect other spins in the molecule.

Additionally, because nuclear spins do not align exactly with the external magnetic field (a result of quantum mechanics described in Chapter 5), they precess about the axis parallel to the external field, a phenomenon known as Larmor precession. A smaller oscillating magnetic field perpendicular to the external field can be applied. When this field is at the precession frequency, then a resonance condition occurs in which the overall magnetization is rotated by 90° in the sample. The frequency of the oscillating magnetic field determines the precession frequency, which depends on the external field. This precession frequency is also affected by the shielding of the external electrons because it depends on the magnetic field, which is adjusted at the nucleus.

As discussed previously, in the presence of an electric field, a second-order effect occurs. The electric field alters the shielding somewhat. In an isotropic solution, this shift is quite small and possibly unmeasurable with current NMR technology. However, the very nature of this work results in molecules that have significant electric dipole moments. Chiral molecules generally have an electric dipole moment, and the electric dipole moments of amino acids are generally relatively large. Thus, chiral molecules align in the external electric field. This has the effect of emphasizing certain components of the shielding tensor in each vector direction as opposed to averaging over off-diagonal components.

With the results of Appendix A, the magnitude of this effect can be estimated in an NMR experiment. Because the nuclear spins do not align with the field, there is a net magnetization in the x-, y-, and z-directions, with the precession about the z-axis. The magnetization then rotates in the xy plane. If a static electric field points in the y-direction, then there is a change in the z-component of the magnetization. This change, from Appendix A, is proportional to the x-component of the magnetization and electric field vector:

$$\Delta M_z \propto M_x E_y.$$

Because the magnetization precesses about the z-axis, then M_x is always changing at the characteristic precession frequency. We thus expect a change in the magnetization in the z-axis, which oscillates at the precession frequency. For a field of ~10 T and an electric field of 10^6 V m^{-1}, the relative shift in the z-component of magnetization relative to the x-component is estimated at

$$\left|\frac{\Delta M_z}{M_x}\right| \approx 10^{-7} - 10^{-9}.$$

This is a small shift, but not necessarily outside the capabilities of modern NMR instruments. The z-component of magnetization shifts in the opposite directions to the L- and D-enantiomers. However, because the x-component of magnetization is precessing, then the z-component oscillates equally for both L- and D-enantiomers.

By inducing a very slow oscillation in the external electric field, one in which the amplitude is measurable in time, the shifts could be measurable (Buckingham and Fischer 2006). For an electric field modulation, this would result in an oscillation of the magnetization in z, which changes with the electric field in the range $\omega \pm f$, where ω is the precession frequency.

This would be a very small chemical shift for each enantiomer but is within the realm of possible measurement. To develop such an experiment, one must create an NMR probe in which an electric field can be induced—perhaps with the sample in a parallel plate capacitor. The electric field necessary is reasonable, though care must be taken to avoid sparks or dielectric breakdown. It would also be helpful to cool the experiment to enhance the signal-to-noise ratio.

In addition, the Buckingham effect also affects the electric dipole moment of the molecule in a magnetic field. Although this does not affect the SNAAP model, a direct result of the Buckingham effect is that the molecular electric polarization is perturbed slightly by the local nuclear magnetization in the molecule (Buckingham & Fischer 2006). By examining the polarization as a function of the magnetization, one may be able to determine the strength of the shielding tensor as well as the chirality of the molecule. This could be done following a $\pi/2$ pulse in the NMR experiment. The transverse magnetization could be measured along with the electric dipole polarization, which could be measured with capacitor plates.

Finally, such experiments do not have to involve amino acids to test the veracity of the calculations, nor do the initial experiments have to examine nitrogen. Shielding of nuclei that are easier to work with would be ideal for initial experiments. Other methods may include doping the amino acids with ^{15}N, which has sharp, but weak resonances. One could also dope compounds with ^{13}C for initial tests of the shielding asymmetry in chiral compounds. NMR science is currently at the crossroads for this sort of work with experimental techniques that utilize electric fields as chiral discriminators proposed (Walls & Harris 2014; Garbacz 2016).

8.3 Astronomical and Space Mission Tests of the SNAAP Model

Although it would surely be helpful if astronomers could identify amino acids from the emission or absorption spectra from space, one look at Figure 4.4 shows how difficult that is going to be. It will be challenging to ever move beyond identification of the spectral lines as being from anything other than polycyclic aromatic hydrocarbons, PAHs. Furthermore, measuring the chirality of the molecules thus identified will present even more of a challenge for radio astronomers. Unless an incredibly intense set of lines from an amino acid, probably not mixed with lines from other amino acids, can be found, this measurement will probably remain on the wish list of astrobiologists for some time to come.

However, the space missions *Hayabusa2* (http://global.jaxa.jp/projects/sat/hayabusa2/) and *OSIRIS-REx* (https://www.nasa.gov/mission_pages/osiris-rex/about) provide considerably more hope that some of the questions associated with the origin of amino acid chirality selection may be answered. Both might be able to obtain material that would enable discrimination between models of chiral selection. The crucial difference between the models that rely on light for chiral selection, specifically the CPL model and the Magnetochiral Anisotropy (MCA) model, and the SNAAP model is that, with neutrinos doing the processing, the amino acids within an entire meteoroid undergo chiral selection, whereas with the other two models, only the surface molecules will be processed.

OSIRIS-REx will use its sampler, the Touch-And-Go Sample Acquisition System, TAGSAM, to collect material from asteroid 101955 Bennu, and return to Earth with it in 2023. TAGSAM will make contact with Bennu via a gentle landing on the asteroid. Following that, TAGSAM will emit a burst of nitrogen gas, which will blow regolith particles smaller than 2 cm into the sampler located at the end of the robotic arm. The hope is for it to collect 60 g of material. Enough gas exists for up to three attempts to collect material. The pads on the end of the sampling head will also collect grains smaller than 1 mm. After collection, the Sample Return Capsule will be opened to allow the sampler head to be stowed, following which the robotic arm will be retracted for *OSIRIS-REx*'s return to Earth.

Of some concern with this mission is whether or not the material collected will be of sufficient size to discriminate between amino acids on the inside and those on the surface. This will depend on the penetration depth of the light that is assumed to do the processing, which depends on the spectrum assumed for the light.

Hayabusa2, however, may be able to provide more definitive information about the processing particles with its sampling of the material from asteroid 162173 Ryugu. Its Small Carry-on Impactor, SCI, is an explosive penetrator, consisting of a copper projectile and a charge. The penetrator and charge will be dropped off, hitting the surface with a velocity of 2 km s^{-1}. *Hayabusa2* will be allowed to orbit around to the other side of the asteroid before the charge is detonated. When the spacecraft orbits around to the location of the crater formed from the impactor and the explosion, it will deploy a lander that will be able to dig the now-softened material at the site of the crater, hopefully obtaining material from well within Ryugu. Some of this material, at least, should have been shielded from processing photons, no matter what their energy spectrum.

Of course, there are uncertainties with this situation also. If Ryugu was formed before the light or neutrinos did their processing, then the answer should settle the question of whether or not neutrinos were involved. If it was formed from agglomerated material from a dust cloud or the accretion disk from the neutron-star–supernova-type systemassumed in the SNAAP model, then the shock wave from the supernova explosion could have compressed preprocessed amino acids into larger objects.

Nonetheless, if the material from below Ryugu's surface is found to have processed amino acids, it will make it easier for the SNAAP model to explain the result and add another layer of complexity to the CPL and MCA models to show how the light they assume to have done the processing might have been able to process material that then was agglomerated into larger objects.

8.4 Future Theoretical Work

The SNAAP model presents one possible scenario for the origins of amino acid homochirality, with concepts rooted in chemistry, biology, physics, astronomy, and geology. Although the model is compelling, there is still much work to be done.

Here, we describe some future work in determining the origins of amino acid chirality.

8.4.1 Autocatalysis

Perhaps a critical piece of any model of amino acid homochirality formation is the process of autocatalysis through which a small enantiomeric excess will evolve to an *ee* of 100%. How a small imbalance becomes a single enantiomer is a question that has been discussed for over 60 years (Frank 1953). Studies have not just concentrated on theoretical models, but experimental studies have also been performed, some of which can produce nearly 100% *ee* with very low initial enantiomeric excesses.

Autocatalysis in the laboratory takes several forms. Chemical autocatalysis relies on reactions to create large *ee*'s (Soai et al. 1995; Soai & Sato 2002; Arseniyadis et al. 2004; Breslow & Levine 2006). Although these studies do support a plausible explanation for enantiomeric excess enhancement, some or many of these experiments do not necessarily apply to prebiotic conditions. However, they could provide a clue for autocatalysis in an early Earth environment. Mechanical enantiomeric amplification may also occur. One example occurs in racemate crystals, which have been shown to selectively precipitate in solution (Klussmann et al. 2006).

Models for autocatalysis have been around for a long time (Blackmond 2010). The initial Frank model (Frank 1953) is a mathematical model in which enantiomers are mutually antagonistic; when they interact with each other, they deactivate while all enantiomers can self-replicate. In this model, even a slight imbalance can result in a large enantiomeric excess.

Analytical chemistry has provided possible solutions to the question of autocatalysis. It was found (Soai et al. 1995) that catalysts in mixtures that are only slightly imbalanced in favor or one enantiomer can drive reactions towards a particular chiral imbalance. This experiment requires an environment that would be unlikely in prebiotic conditions, though it does provide a starting point. Other nonchemical models may be possible as well. These include crystal formation in solutions, thermodynamic partition models, and models of molecular energy states.

Needless to say, models of autocatalysis will likely continue to attract the efforts of astrobiologists and will certainly continue to be a compelling field of study in chemistry.

8.4.2 Environmental Effects

The SNAAP model assumes that the interactions involved take place within meteoroids. However, the interiors of meteoroids are varied and complicated in some cases. Amino acids trapped in meteoroid or even cometary material may be in an aqueous environment, or the environment may become aqueous later on. There appears to be a close relationship between the aqueous alteration that has occurred within a meteoroid and the level of enantiomeric excess of the amino acids within the body (Glavin & Dworkin 2009). This result alone provides evidence of the variation in the characteristics of meteoroids that contain amino acids. Depending on the composition of the meteoroids that delivered the amino acids to Earth, there could be several possibilities for the environment within the meteoroid.

In addition, the aqueous nature of the meteoric environment will determine the molecular geometry of the amino acid, as shown in Chapter 7. An aqueous environment will tend to result in zwitterionic amino acids. Further, the pH of the environment can determine if the amino acid is cationic or anionic. As was shown in Chapter 7, the charged nature of amino acids is directly related to the anisotropy of the shielding tensor.

While examining the environmental effects could be potentially complicated, it would be interesting to model these effects based on the specific conditions of meteorites that have delivered their cargo to Earth.

8.4.3 Astrodynamics

So far, the basic tenets of the SNAAP model have been presented. Amino acids in crossed electric and magnetic fields may undergo interactions with neutrinos that selectively destroy one chiral state over another.

However, there are many ways electric and magnetic fields may be generated. The motion of a meteoroid in an external magnetic field can cause an electric field in the rest frame of the amino acid. However, the meteoroid does not have to be moving much at all for an electric field to be formed. If the source of the magnetic field is rotating about an axis that is not the field axis, then the field is changing in the space surrounding it. Meteoroids trapped in the vicinity of the field will have an electric field, which is related to the magnetic field by

$$E = \frac{-dA}{dt},$$

where the vector field $\boldsymbol{A}$ is the vector potential, which describes the magnetic field in all of space. It is the magnetic equivalent of the electrostatic (or electric) potential, which relates the potential (in volts) to the electric field (in V m^{-1}). The vector potential is related to the magnetic field by

$$\boldsymbol{B} = \nabla \times \boldsymbol{A},$$

It is quite possible to have a situation in which the magnetic field is moving with respect to the space around it. Neutron stars can rotate rapidly, creating large changes in the local magnetic field. Pulsars result from the interactions of rotating neutron stars and their magnetic fields.

Besides this, the magnetic field gradient near a neutron star can be as high as 10^4 T km^{-1}. This is enormous. Even slowly moving meteoroids could have significant electric fields in such a region of space.

Of course, it seems unlikely that a meteoroid would be stationary near a neutron star. It should either orbit the star, pass by it, fall into it, or acquire enough energy to be ejected from the system. In the case of the binary system discussed in Chapter 7, the orbital mechanics can be quite complicated. Objects that enter the system from the outside may undergo very convoluted orbits before being ejected from the system. Objects in the accretion disk may move much more slowly than they otherwise might due to disk viscosity. If objects are formed as part of the planetary system, it may be possible for them to be trapped at some of the more stable Lagrange points in the system.

All this is to say that the motion of meteoroids in a binary system, which may also have an accretion disk, can be quite complicated. In a binary system with no accretion disk, the orbital mechanics can be solved using Newtonian equations, though general relativistic effects should be accounted for near the neutron star. In binary systems with smaller orbital objects, gravitational force is not the only effect to be reckoned with. Centrifugal effects will also change the orbital mechanics of the system. When an accretion disk is added, fluid mechanics computations are necessary (Popham & Sunyaev 2001).

One can imagine the enormous number of possibilities for a neutron star, WR star, and meteoroid orbits, velocities, positions, fields, masses, and accretion rates. Given this, evaluating the effects of the SNAAP model and optimum conditions for the SNAAP model to occur is a formidable task. However, further evaluation and precision predictions of the SNAAP model will require modeling of the astrodynamics involved.

8.4.4 Molecular Chemistry

The SNAAP model does not occur in a vacuum. If the SNAAP model is fast, then the effects of molecular chemistry may be negligible. However, if the SNAAP model occurs over several years, decades, or millennia, then the effects of chemistry could be significant.

One possible effect would be that amino acids may be forming while some of them are being destroyed by neutrinos. This could have a positive effect, because it provides more fuel for the SNAAP model to process. Even more, autocatalysis mechanisms may take advantage of the additional amino acids.

Another potentially interesting chemical effect is that of ^{14}C. In the SNAAP model, ^{14}N is converted to ^{14}C, which is radioactive and turns back into ^{14}N via the reverse reaction to antineutrino capture:

$$^{14}\mathrm{C} \rightarrow {}^{14}\mathrm{N} + e^{-} + \overline{\nu}_e.$$

This reaction has a half-life of about 5700 years. Depending on the site of the SNAAP model, these decays could provide a supply of nitrogen for the subsequent amino acid formation, the chemistry involved in autocatalysis, or the formation of nitrogen compounds that change the environmental chemistry. Such a complex network of possibilities would be interesting to explore.

8.5 Effects on the Rest of the Molecule

What about neutrino effects on other nuclei within the molecule. Except for cysteine, which contains one sulfur atom, the amino acids contain exclusively hydrogen, nitrogen, oxygen, and carbon. However, as was stated previously, the carbon and oxygen do not contribute to chiral selection in the SNAAP model because they have no spin. In the case of cysteine, sulfur also has no spin. It could be possible that hydrogen contributes to chiral selection because it has a spin of 1/2. However, in this case, the spin-aligned state and the spin-antialigned states would have the same cross section. The spin-aligned state would have a total angular momentum of 1, while the spin-antialigned state would have a total angular momentum of 0. In either case, the electron or antineutrino would not have to provide a unit of angular momentum as it would be emitted with its spin in such a way as to form the total angular momentum of the molecule.

Hydrogen also has a very small nuclear magnetic polarizability compared to nitrogen. Nitrogen is buried in the molecule, while all of the hydrogen are near the edges. In this case, the differences between L- and D-nantiomers are very small compared to the shielding tensors of nitrogen.

Neutrinos could still destroy these molecules, but we expect them to be destroyed in equal amounts for both L- and D-states. This would certainly contribute to the total number of molecules, but not to the enantiomeric excess. Certainly, this is something that can be explored as a contribution to the total model, however.

Although we do not expect the ^{14}N magnetic substate with zero projection along the magnetic field axis to have much of an effect on the *ee*'s that are obtained in the SNAAP model, it will contribute at least to a slight diminution of the effects from the projection $+1$ and -1 substates. Although the contributions from the zero substate have not been included in the work we have done thus far, it will have some effect on the final results, and therefore does need to be included for a complete picture.

8.5.1 Galactic Molecular Evolution Models

In astrophysics, the field of galactic chemical evolution (GCE) is reserved for the predictions of the formation and transport of the elements through the galaxy to form the stars seen today, including our own solar system. The first few chapters of this book relate the formation of the elements to what is observed in the galaxy today. The earliest stages of element production began with the Big Bang, followed closely by the deaths of the most massive stars in the universe, and later on by neutron star mergers, Type Ia supernovae, and a few other nucleosynthesis events. GCE continues today with the same events. Cosmic rays provide a continuous process that is always adjusting the elemental abundances in our galaxy.

However, these elements eventually form molecules, organic molecules, and even complex organic molecules. The study of the evolution of organic matter in space is closely related to the fundamental study of the origin of matter (Ehrenfreund et al. 2011).

Much like GCE, the study of galactic molecular evolution requires a production mechanism by which molecules can be created, a destruction mechanism by which molecules are destroyed, and a diffusion mechanism by which molecules are transported through space (Snow & McCall 2006). Results from models of these studies can be compared to radio and IR astronomy observations (Liszt & Lucas 2000) as well as measurements of meteoric content (Cronin & Pizzarello 1997).

Models of molecular evolution can be used to link the formation and delivery of molecules to the site or sites in which they were produced. Questions that can be answered (or posed) include whether Earth's amino acids have their origins in a single event; whether organic molecules been delivered to other parts of the galaxy, and if so, whether they formed the same way; and a host of other queries.

Much of the modeling for large-scale, long-term determinations of organic molecular evolution can follow—at least on a basic scale—the same type of modeling used for GCE. Many of the chemical reaction rates have some of the same dependences as the nuclear reaction rates. There may be additional complications in developing an evolution model to include UV and X-ray exposures, which generally do not affect nuclear reaction rates. Additionally, while the autocatalysis mechanism is inferred, it may not be well understood. However, such modeling might constrain or hypothesize autocatalysis rates and times.

8.5.2 The Formation of Sugars

Finally, it is worth noting that—while amino acids are left-handed—sugars are right-handed in structure. This is closely related to the origins of biomolecular homochirality. Will the SNAAP model also produce right-handed sugar molecules? Sugars may very well have an origin comparable to that of amino acids. Alternatively, the origin of complex carbohydrates may be terrestrial (though simple sugars have been found in space; Sewiło et al. 2018). Although the formation of sugars may be linked to the formation of amino acids, it may also be completely independent.

References

An, F. 2017, EPJ Web Conf., 164, 07002

Arseniyadis, S., Valleix, A., Wagner, A. & Mioskowski, C. 2004, Angew. Chem., 43, 3314

Avignone, F. & Efremenko, Y. 2003, JPhG, 29, 2615

Balantekin, A. B. 2006, hep-ph/0601113v12006hep.ph, 1113B

Bird, M. D., Bole, S., Eyssa, Y. M. & Gan, Z. 2000, ITAS, 10, 451

Blackmond, D. G. 2010, CSH Perspect. Biol., 2, a002147

Bradt, J., Bazin, D., Abu-Nimeh, F., et al. 2017, NIMPA, 875, 65

Breslow, R. & Levine, M. S. 2006, PNAS, 103, 12979

Brumfiel, G. 2008, Natur, 455, 436

Buckingham, A. D. 2004, CPL, 398, 1

Buckingham, A. D. & Fischer, P. 2006, ChPh, 324, 111

Cronin, J. R. & Pizzarello, S. 1997, Sci, 275, 951

Ehrenfreund, P., Spaans, M. & Holm, N. G. 2011, RSPTA, 369, 538

Excoffier, S., Bertin, P. Y., Brossard, M., et al. 2005, NIMPA, 551, 563

Frank, F. C. 1953, BBA, 11, 459

Garbacz, P. 2016, JChPh, 145, 224202

Glavin, D. P. & Dworkin, J. P. 2009, PNAS, 106, 5487

Klussmann, M., Iwamura, H., Mathew, S. P., et al. 2006, Natur, 441, 621

Kobayashi, T., Chiga, N., Isobe, T., et al. 2013, NIMPB, 317, 294

Liszt, H. & Lucas, R. 2000, A&A, 355, 333

Myrgorodska, I., Meinert, C., Martins, Z., d'Hendecourt, L le & Meierhenrich, U. J., S. 2016, JChA, 1433, 131
Plies, E., Marianowski, K. & Ohnweiler, T. 2011, NIMPA, 645, 7
Popham, R. & Sunyaev, R. 2001, ApJ, 547, 355
Puglisi, M., Stipcich, S. & Torelli, G., ed. 2012, New Techniques for Future Accelerators II: RF and Microwave Systems (Berlin: Springer)
Sewiło, M, Indebetouw, R., Charnley, S. B., et al. 2018, ApJL, 853, 19
Shane, R., McIntosh, A. B., Isobe, T., et al. 2015, NIMPA, 784, 513
Snow, T. P. & McCall, B. J. 2006, ARA&A, 44, 367
Soai, K., Kawasaki, T. & Matsumoto, A. 2014, Chem. Rec., 14, 70
Soai, K. & Sato, I. 2002, Chirality, 14, 548
Soai, K., Shibata, T., Morioka, H. & Choji, K. 1995, Natur, 378, 767
Tang, L., Yan, C. & Hungerford, E. V. 1995, NIMPA, 366, 259
Walls, J. D. & Harris, R. A. 2014, JChPh, 140, 234201

Richard N Boyd and Michael A Famiano

Chapter 9

How Have Scientists Previously Explained the Amino Acid Chirality?

9.1 Introduction to Models

The explanations for amino acid chirality falls into two categories, and they are not necessarily distinct. These are the deterministic and stochastic models. A diagram showing how these models are related and a few of the mechanisms for each is given in Figure 9.1.

Deterministic models are those in which an external mechanism drives amino acid formation or destruction toward a particular chirality and has the same result for the same conditions everywhere. With a deterministic model, all amino acids everywhere are left-handed (at least for the same environmental conditions). The SNAAP model draws elements from a few of these mechanisms as shown in Figure 9.1.

Stochastic models are those in which amino acid chiral selection is due to a random or statistical process. If the experiment were repeated, a different result may appear. Stochastic models could predict that there are planets out there with right-handed amino acids (or maybe both).

It appears that no mechanism that produces amino acid chirality, with the possible exception of the SNAAP model, can generate very much enantiomerism, that is, a large imbalance toward one chirality. Most models require that some mechanism or mechanisms that can produce some enantiomerism exist, and that at least one other mechanism that can amplify that enantiomerism to at least the levels found in the meteorites, and ultimately to homochirality—the existence of a single chiral form—must also exist.

There have been many explanations for the origin of amino acid chirality. For those astrobiologists whose models we do not discuss, but which may be superior to the ones we do discuss here, we apologize for overlooking your model and invite an offline discussion. We will try to summarize some of them, especially the ones we think are coming fairly close to producing feasible explanations. We classify models

doi:10.1088/978-0-7503-1993-5ch9

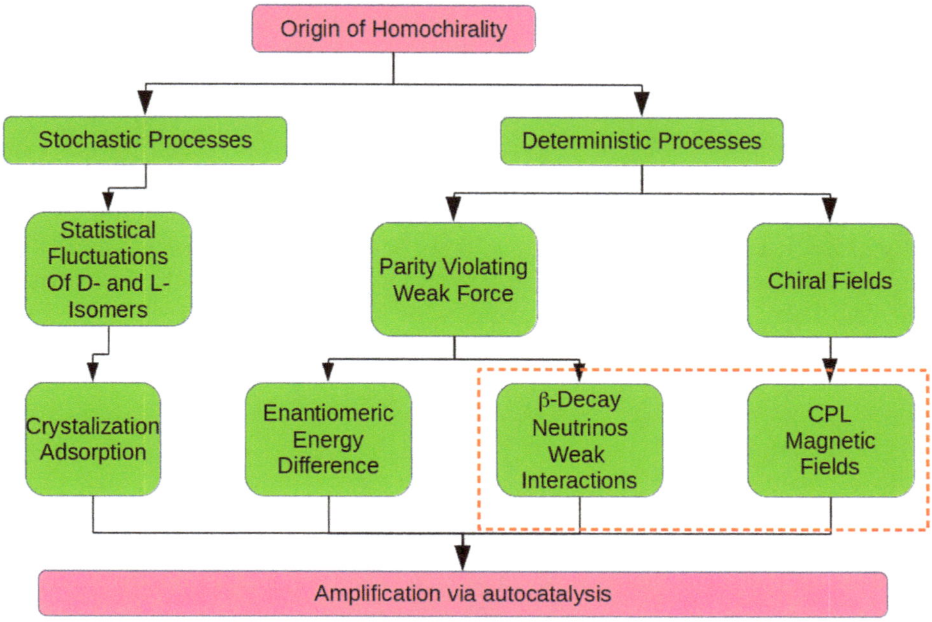

Figure 9.1. Diagram of the classification of models of amino acid chiral selection. The SNAAP model combines elements of the weak interaction and magnetic fields. These elements are indicated with the red square in the figure.

of amino acid homochirality into two classes: those claiming that amino acids have a terrestrial origin and those that claim they originated in outer space. Those in the former category still have to account for the chiral amino acids found in meteorites.

9.2 Models that Produce Chirality

The suggestion that amino acids were created by a lightning bolt in an appropriate Earthly chemical environment was certainly given support by the Miller–Urey (Miller et al. 1959) experiment, discussed in previous chapters. This scenario, the Earthly creation of amino acids, was reviewed extensively by Wachterhauser (1988, 1992). Of course, the chirality issue presents a problem that this scenario must address. Thus, there have been many attempts to explain how the equal populations of left- and right-handed amino acids that would have existed initially could have been converted to one that is entirely left-handed.

9.2.1 Chiral Selection via Circularly Polarized Sunlight

One possibility that has been suggested to make the Miller–Urey model consistent with Earth's amino acid chirality is to assume that CPL from our Sun might create some amount of enantiomerism as it impinges on the amino acids, since one chirality of a molecule would then be more readily destroyed than the other (Bailey et al. 1998; Takano et al. 2007). This could happen under the following circumstances. A tiny fraction of the light from the Sun is polarized (less than 1%). It is usually assumed that the light has to scatter twice in order to achieve circular polarization, although we will discuss a model below where that is not the case. We will assume that CPL is created when the light scatters from dust grains. The dust grains have to be nonspherical, in which case the interstellar magnetic fields (which are tiny but nonzero) can provide a tiny bit of orientation for them on average. Then, it can be shown that a single scattering will produce linearly polarized light (see Chapter 5). However, the orientation of the interstellar magnetic fields is pretty random, so if the light scatters a second time, it will most likely be from a grain, again assumed to be nonspherical, that is oriented differently from the grain on which the first scattering occurred. Then, this second scattering will change the phase of one of the components of the linearly polarized light, converting the linearly polarized light into circularly, or perhaps elliptically, polarized light. Since the ability of CPL to produce enantiomerism has been demonstrated in the laboratory (Modica et al. 2014), it should also work in nature.

The problem with the above model is the randomness of the magnetic field directions in interstellar space. Thus, the CPL produced by the two scatterings is just as likely to create right CPL as it is to create left CPL. If the CPL creates left-handed molecules in one place, it will be circularly polarized in the opposite direction somewhere else, and so it will make right-handed molecules there. The fractions of the two chiralities averaged over all space will be expected to be equal. Sunlight has an additional interesting feature: its CPL varies during the day, and the different components during the day will just cancel each other. The attempts to explain how this local chirality but global equality of right- and left-handed molecules can be circumvented for sunlight border on the heroic; they are discussed in the review paper by Bonner (1991). After thoroughly discussing them, he professes skepticism about their probability for success. Bailey (2001) comes to the same conclusion.

The other possibility is that the amino acids were created in the cosmos, then transported to Earth by meteoroids or comets, or were transported here by an advanced civilization. Front and center in this aspect of the discussion is the Panspermia hypothesis: that life was not created on Earth, but came to Earth from outer space. This was mentioned as early as the fifth-century BC by the Greek philosopher Anaxagoras (O'Leary 2008), and it has reemerged many times since then. Its proponents have included Kelvin, von Helmholtz, and Arrhenius, and in more recent times, Fred Hoyle, Chandra Wickramasinghe, and Francis Crick. Many of the Panspermia explanations have little in common with the original one except that they all have the origins of the molecules of life external to Earth. The purest version of Panspermia purports that life, and hence chirality, has always existed; this would certainly force a revision of our current understanding of the Big Bang, since it is difficult to imagine molecules at the temperatures that existed in the early universe—certainly not during the hundreds of millions Kelvin that existed at the time when nucleosynthesis occurred a few minutes after the Big Bang, which are even considerably higher than those prior to Big Bang Nucleosynthesis.

Nonetheless, the suggestion that the seeds of life did not originate on Earth but rather arrived in some way from outer space is certainly not new, and it was given a huge boost by the results of the analyses of the Murchison meteorite, as discussed previously. It showed that amino acids and other biologically interesting molecules are produced in outer space, that some of the amino acids have been chirally selected to be left-handed, and that they can survive the journey to Earth. Panspermia lives!

However, as often happens in science, this simply moves the question to be addressed to a higher, or in this case a more distant, level of sophistication. The nagging question of the origin of the chirality of amino acids still remains to be answered and understood. The nonzero chiralities of the amino acids from the Murchison meteorite and other meteorites were predominantly left-handed, and the amino acids on which we depend for life are exclusively left-handed. But is left-handedness inevitable for amino acids? Is there some mechanism by which they are produced that always ultimately makes them left-handed? Or are they produced as racemic collections of molecules and then processed to make them enantiomeric? Can we use the chirality of amino acids to understand their origins and possibly even our origins? Indeed, is understanding the origin of the chirality of amino acids crucial to understanding the origin of life?

9.2.2 Chiral Selection via Starlight

The possibility of producing left-handed amino acids via CPL can certainly be applied to the interstellar medium as well as on Earth. Indeed, photons may be more effective if they are at energies higher than solar photons, which must penetrate Earth's atmosphere to get to its surface and so will only have the energies of visible light photons. Thus, ultraviolet photons, which cannot penetrate Earth's atmosphere but are plentiful in outer space, may be effective in processing amino acids in space. Indeed, it has been demonstrated in the laboratory that this effect can occur (Modica et al. 2014).

It should also be noted that Earth's atmosphere developed after Earth was formed and well after life is thought to have begun. Thus, ultraviolet radiation could have had an effect on processing the amino acids that existed on Earth early in its history (although Earth would still have had some sort of atmosphere, possibly the result of volcanic explosions and dust from huge meteorite impacts). The polarization appears to be highest when photons have to pass through a relatively dense nebula in order to escape their star, which gives them more opportunities to scatter. However, starlight has been observed to be circularly polarized at more than a fraction of a percent in only a few exceptional cases. So, this scenario is roughly as promising for ultraviolet photons in outer space processing the amino acids as it is for sunlight affecting them on Earth. We will return to this type of model later.

As in the model where amino acids are processed on Earth by sunlight, CPL would destroy molecules of one chirality but not affect the other, or more likely destroy both to some extent, but would affect one chirality more than the other.

Bailey (2001) discusses the possible sources of CPL in a model where amino acid chirality might have resulted from ultraviolet photons in space, and this model has been developed in detail by many others (Modica et al. 2014; Bailey 2001; Gledhill & McCall 2000; Hakala et al. 1994; Ehrenfreund et al. 2001; deMarcellus et al. 2011). Bailey presents a table that shows the amount of amino acid destruction that would be required to produce different enantiomeric excesses. If it is assumed that the light is 100% circularly polarized, destruction of 99.6% of the molecules would be required to produce an enantiomeric excess of 10%, and of 99.975% to produce 15%. If the light is only 1% circularly polarized, achieving an enantiomeric excess of 10% would require destruction of 99.996% of the molecules. This puts a high premium on performing the chirality selection with light that is as highly circularly polarized as possible.

The model developed by Gledhill & McCall (2000) describes the polarization that would be realized from the single scattering of ultraviolet light from nonspherical grains that were aligned by a magnetic field. They found that grains having a longer-to-shorter-dimension ratio of 2:1 could produce circular polarizations as high as 50%. Objects with circular polarizations approaching this value have been observed (O'Leary 2008), suggesting that objects that have produced such magnetic fields over a volume with an abundance of grains may exist.

Bailey (2001) also discusses the various possible sources of CPL that have been explored in the literature. This includes synchrotron radiation produced by relativistic electrons being accelerated in the magnetic field of a neutron star. However, such light has not been observed to be polarized and so must have an extremely small polarization. It also includes CPL from a white dwarf that is accreting matter from a companion star. This could produce an extremely high circular polarization, but Bailey argues that the probability that the material making up our solar system would have been subjected to such a source is small. However, he also notes that if life in the universe is found to be rare, this might be a plausible source.

Bailey concludes that reflection nebulae would seem to be the most promising source of processed amino acids, especially given the model of Gledhill and McCall for producing CPL. In contrast to the situation with the accreting white dwarf, Bailey notes that if life in the universe is not rare, then this source would seem to be a major contributor. He also assumes that the processing would occur in a relatively dense cloud, so that the processed molecules would only undergo one processing event and would not even see photons from other stars.

However, this cosmic CPL scenario has the same difficulty producing global amino acid chirality as the circularly polarized sunlight scenario, that is, CPL that was left-handed in the cosmos would be expected to be balanced by right-handed CPL. So, this model might predict that there would be regions of the interstellar medium where amino acids would be left-handed, and others where they would be right-handed, just as would be the case for the sunlight scenario. This assumes that the interstellar magnetic fields are not fluctuating, which might be expected to be the case for the starlight passing through the nebula produced by its star of origin either from a stellar explosion or from stellar winds. But averaging over all space would make it difficult to produce anything other than a racemic result. This does remain a general problem for the CPL explanation in both the terrestrial origin and the extraterrestrial origin models as long as left-handed amino acids are all that we find. Note, though, that our solar system might be smaller than the typical regions that are processed, so that regions with right-handed amino acids may exist, but far enough from us that we would not be able to identify them. We return to this issue below.

Another problem resides with the larger objects that might be processed by CPL. The light would only process the surface of the object, so if it were a large meteoroid or a comet, the small fraction of the chirally selected molecules on the surface would translate into an infinitesimal enantiomeric excess when the entire volume of the object, presumably with many more racemic molecules inside the surfaces, is taken into account. And, since small objects would tend to be racemized or even destroyed when they encountered Earth's atmosphere, it is presumably the large objects that delivered Earth's preferred chirality.

This problem could be circumvented by assuming that molecules could be processed in space or possibly on the surfaces of small dust grains. The dust grains would gradually agglomerate and ultimately form much larger objects such as meteoroids. This would produce objects that would not necessarily be uniformly chiral throughout, but would be large enough for a significant fraction of each one to pass through the atmosphere of any planet on which they ultimately came to reside.

However, timing would be critical to this scenario, since the half-lives estimated for amino acids in a high photon flux environment are not long—possibly as short as several hundred years (Ehrenfreund et al. 2001)—so that grain formation must occur on a sufficiently rapid timescale to shield the processed molecules that had been selected. Once they have agglomerated into a larger clump of material, their half-life increases to millions of years. However, grain clumping must not occur too rapidly, as the processing to establish chirality must occur on a shorter timescale.

An interesting approach described by deMarcellus et al. (2011) assumes that the regions that are bathed in fairly uniform CPL are large enough to encompass entire planetary systems. Technically, we have no data that tell us that the galaxy's amino acids are all left-handed, and currently the data all originate from within our solar system (including those from comets). Thus, it is possible that the star, or stellar region, that produced the CPL that made our solar system with left-handed amino acids produced oppositely directed CPL that made another stellar system somewhere else with right-handed amino acids. The global universal chirality would be zero, but we would not know, not yet anyway, about the right-handed planetary system. This model assumes that the processed regions did not get mixed, or there would be pockets of left-handed and pockets of right-handed amino acids. Of course, those might also exist unbeknown to us.

A number of different regions of space with light having a large enough circular polarization, at least over some bandwidth, have been identified as possibilities where the necessary amino acid processing can occur. Interstellar CPL is thought to have a short wavelength cutoff as well, limiting the bandwidth to the regions necessary for amino acid processing (Takano et al. 2007). Recent studies have found large areas of CPL in star-forming regions (Bailey et al. 1998; Fukue et al. 2010). This situation is sketched in Figure 9.2.

Experiments have been performed (Modica et al. 2014; Hakala et al. 1994; Ehrenfreund et al. 2001; deMarcellus et al. 2011) with beams of ultraviolet CPL from synchrotron radiation facilities in France and Denmark to measure the

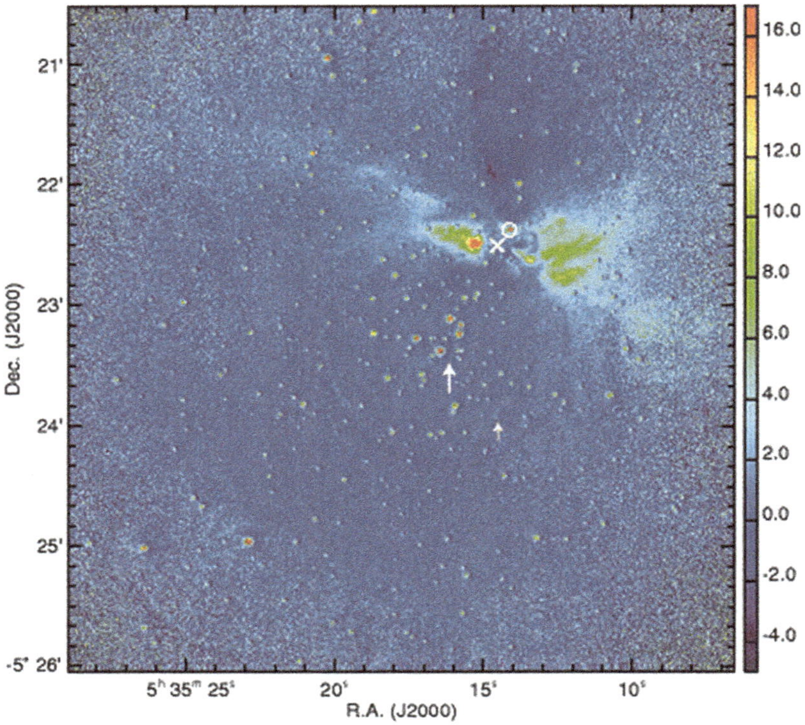

Figure 9.2. Image of degree of circular polarization (%) in the *K*-band (2.14 μm) of the central region of the Orion star-forming region. The field of view is 5.5 arcminutes or 0.74 parsec^2 at a distance of 460 parsec (one parsec = 3.086×10^{13} km). North is up and east is to the left. The positions of IRc2 and BN are indicated by a cross and a circle, respectively, while those of the Trapezium stars and the low-mass young star OMC-1S are denoted by big and small arrows, respectively. A positive sign for CP indicates that the electric vector is rotated anticlockwise in a fixed plane relative to the observer. From Fukue et al. (2010), courtesy of Springer Nature.

likelihood of producing enantiomerism in amino acids. The beams had a well-defined wavelength, and so produce a detailed measurement of the effects of CPL as a function of wavelength. Enantiomeric excesses as high as a few percent (Meinert et al. 2012) were obtained at specific wavelengths, as is indicated in Figure 9.3. However, both positive and negative *ee*'s were produced, at slightly different wavelengths, and they average out to a much lower *ee*, usually at wavelengths that are fairly close together. This is as expected from the Kuhn–Condon zero-sum rule.

This could be a problem for the CPL model, at least if it relies on scattered photons or photons that are emitted from a hot star. The spectrum for such photons would be much broader than the typical separation between the positive and negative *ee* wavelengths, as is illustrated by the blackbody spectra indicated in Figure 9.3, and the scattering would be expected to broaden even further the spectrum of the circularly polarized photons needed to produce the chiral selection. Thus, it would be difficult to avoid having the positive and negative *ee* points

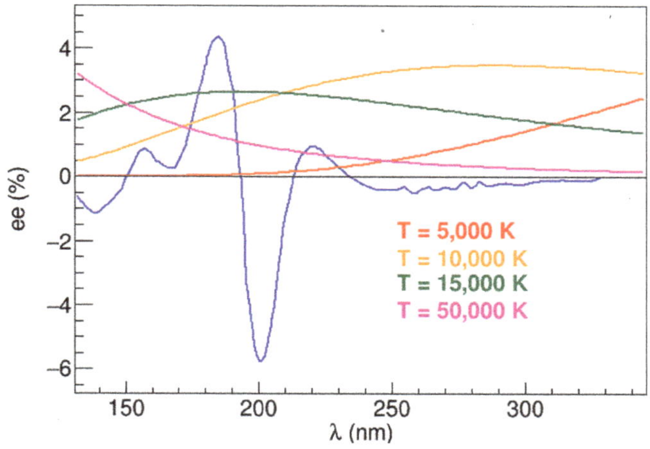

Figure 9.3. Enantiomeric excess as a function of wavelength for alanine, indicated in blue. Superimposed on the figure is a blackbody radiation spectrum expected from stars having a range of temperatures. Note that the light that would produce positive enantiomerism in amino acids in the CPL model would have come from an extremely hot star (the 50,000 K curve seems to work best) and would have undergone scattering, and so would be expected to be even broader than the blackbody spectrum. Thus, the spectrum to which the amino acids were subjected would be even flatter than the blackbody curve. Most of the blackbody spectrum shown for the 50,000 K temperature is off the graph to the left and vastly exceeds the scale of the ordinate, and so would destroy a much larger fraction of molecules than could have been made positively enantiomeric. Data are from Meinert et al. (2012).

average each other out to a net *ee* of far less than a percent, unless a source with much better defined wavelengths can be found. Could the light from some cosmic source be so well tuned?

Furthermore, assuming that the low-energy tail of the distribution of the ultraviolet CPL photons is necessary to do the processing, as mentioned above, the light will destroy most molecules of life in order to create a preferred chirality in a few of them.

9.2.3. Chiral Selection via the Weak Interaction

It might be thought that chirality could be achieved by a shift in the energetics of the molecules of opposite handedness, that is, if one of them might be selectively formed if it were more tightly bound than the other. In that case, thermal equilibrium would favor the more tightly bound molecule, which would then end up with the larger abundance. In several publications (Mason & Tranter 1983, 1984, 1985; Mason 1984; Tranter 1985a, 1985b, 1985c, 1986, 1987), Mason and Tranter studied the possible effects of the weak interaction (the one that mediates the β-decay we discussed in Chapter 3) in this regard. They found that the effect was extraordinarily small, about one part in 10^{17} (one part in one hundred million billion). Gol'danskii & Kuz'min (1989) concluded that this effect is so small that weak neutral currents could not be the basis for a "simple evolutionary [thermal] hypothesis." (However, note the discussion by Rode et al. 2007 on the possible role of copper in peptide formation, and its potential role in establishing chirality via

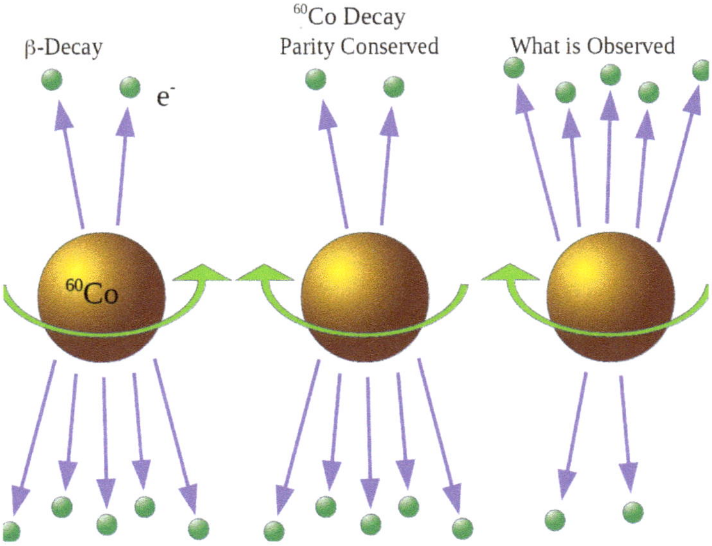

Figure 9.4. The parity nonconservation experiment of Wu et al. (1957). If parity were conserved, the yield would not depend on the direction of the magnetic field, hence of the nuclear polarization. However, that is not the case, as shown on the right. Thus, parity is not conserved. Reprinted by permission from L.D. Barron, "Chemistry: Chirality, magnetism and light," Springer/Nature, 2000, 405, 895–896, doi:10.1038/35016183. Copyright 2000.

the enhancement it would produce on the energy separation of the chiral states.) Nonetheless, some evidence exists that electroweak parity-violating energy shifts could produce an enantiomeric excess if the molecules are in a gas phase (MacDermott et al. 2009).

However, the weak interaction is the one known interaction that, because of its very nature, could produce molecular chirality because it violates parity conservation, which dictates that the mirror image of an object looks the same as the object itself. This means that the weak interaction could produce a chirality simply as a result of the asymmetry that results from the interactions it might produce between particles. This property of the weak interaction was originally suggested by Lee & Yang (1956) and confirmed by the Nobel-prize-winning experiment of Wu et al. (1957). That experiment showed that electrons emitted in the β-decay of ^{60}Co (cobalt) nuclei that had their spins polarized in a strong magnetic field were preferentially emitted in the opposite direction to the spin, while the predictions of parity conservation would be that equal numbers would be emitted in the direction of the spin and in the opposite direction. The weak interaction is the only one of the basic interactions in which parity is violated at the most fundamental level.

The results of the Wu et al. (1957) experiment are given in Figure 9.4, which indicates the rate at which ^{60}Co nuclear decays were detected, in one case with the magnetic field (H) up and the other with it down. The decay electrons could only be detected in the up direction because of the way the experiment was configured. The magnetic field provided the orientation of the spins of the ^{60}Co nuclei, but only when the sample was cooled to near absolute zero. As time progressed and the sample warmed up, the orientation went to zero because of the temperature dependence of the spin state populations, and the count rates of the two magnetic field orientations became the same. The data clearly show that there are significantly more decays resulting from the field down configuration than from the field up configuration when the nuclei are oriented. If parity were conserved in β-decays, the rate at which electrons were emitted would not have depended on the nuclear orientation, and the event rates would not have depended on the magnetic field orientation. Thus, it must be concluded that parity is not conserved in β-decay or in the weak interaction, which mediates β-decay.

Because this interaction violates parity conservation, it might be thought that it could produce an effect that could skew the balance between left- and right-handed molecules. Indeed, since this is the only interaction that is known to violate parity, it might be thought to be the obvious candidate to perform this function. One possible means by which this could come about originated with Vester et al. (1959) and Ulbright & Vester (1962), who noted that the electrons produced in β-decay were longitudinally polarized (that is, the electron's spin is oriented antiparallel to its direction of motion) if they were sufficiently energetic. However, more energetic electrons tend to be more fully polarized. Highly relativistic electrons are nearly completely longitudinally polarized. In interacting with matter, these electrons would produce circularly polarized bremsstrahlung photons, literally, braking radiation. This effect is produced when a charged particle (in this case an electron) is accelerated or decelerated when it encounters another charged particle. The bremsstrahlung photons, in turn, could interact with the molecules to produce chiral selection.

The details of this scenario were studied by Al Mann and Henry Primakov (Mann & Primikov 1981). They attempted to show that the β-decay of ^{14}C, which despite being radioactive is a relatively abundant isotope of carbon because it is produced in Earth's atmosphere, could perform the chiral selection. They were able to show that this interaction could produce chiral selectivity, but only if there was a difference in the destruction rate of left- and right-handed molecules by the electrons. Unfortunately, it was found that the expected effect is extremely small.

Several experiments (Akaboshi et al. 1978, 1981, 1982; Conte 1985) designed to demonstrate this effect produced results that suggested that some chiral selectivity did occur, but it was apparently due to the longitudinally polarized electrons and not the bremsstrahlung radiation they produced. It might seem that this would work as well as the bremsstrahlung photons, but apparently the effects are still extremely small—barely at the level of detectability. Other attempts to confirm the existence of this effect have been unsuccessful, presumably because of the extremely low level of enantiomerism that would be expected from this mechanism (Bonner 2000).

This effect was also studied theoretically by Gol'danskii & Kuz'min (1989). They concurred with Mann & Primakov (1981) that the effect was small, but they actually gave a definitive estimate: it is expected to be less than one part in 10^{10}. This was an upper limit, since they assumed that the electrons were totally polarized, and this would not likely be the case. They also assumed the electron energy to be relatively low, where the probability for producing chirality is higher. But low-energy electrons are less aligned than they would be at higher energies. The actual limit of the effect could easily be lower than was estimated by one or more orders of magnitude, and that would be consistent with the attempts to measure it. Bonner (2000) summarized the situation up to the time that article was written by concluding that parity violation in biomolecules was not likely to be the result of parity violation in the weak interaction.

However, recent work using electrons from radioactive sources to bombard samples of chemicals such as CH_3OH, NH_3, and water (H_2O) mixed with fairly large amounts of metals such as cobalt and copper have produced amino acids that were found to have tiny, but nonzero, enantiomeric excesses (Gusev et al. 2008; Burkov et al. 2008). Although the source of the enantiomerism produced in these experiments is not entirely clear, this is an interesting result.

A related suggestion was made by David Cline (2005), who attempted to advance the idea of Vester and Ulbright. He noted that the incredibly intense neutrino flux from a core-collapse supernova might replace the electrons as the particles that could perform a chiral selection of the molecules with which they might interact. Cline focused on the number of reactions that the neutrinos, specifically, electron antineutrinos, would have with nuclei such as hydrogen. The relevant reaction is $\bar{\nu}_e + {}^1\mathrm{H} \rightarrow e^+ +$ neutron. No corresponding (charged current) reaction can occur for electron neutrinos, since neutrons are unstable particles and thus are not present in isolation. Cline also looked at the effects that the positrons, produced by the electron antineutrinos interacting with the protons, would have on the nuclei. However, he did not describe how this effect could produce molecular chirality. He also apparently did not consider what regions of space would be precluded from molecular interactions by the size of the progenitor of the core-collapse supernova or by the photons produced in the supernova event. We have discussed previously that these effects must be taken into account if a model is to be convincing.

A similar suggestion (Barqueno & Perez de Tudela 2007) involved the interaction of supernova electron neutrinos and antineutrinos with atomic electrons. The interaction between the neutrinos and antineutrinos and the electrons would shift the energies associated with molecules of different chirality by a tiny amount, but by an amount that is comparable to the energies associated with thermal motion in the interstellar medium. However, this depends on the difference in the fluxes of the electron neutrinos and antineutrinos, which is not expected to be large and may be essentially zero. But if thermal equilibrium had enough time to prevail and the energy difference was appreciable, this could (in principle) produce a preference for one molecular chirality over the other. However, as noted earlier, the intense neutrino flux from a supernova lasts only a few seconds, and it is not clear that thermal equilibrium would have time to occur on that timescale in the cold environs of space. This is also in addition to the requisite assumption of an appreciable difference in the numbers of electron neutrinos and antineutrinos. Finally, as noted earlier, the molecules would need to be close enough to the supernova that the antineutrino flux would be high, and that would not be possible unless they were inside the star. We believe that this scenario would be difficult to justify.

9.2.4 The Magnetochiral Anisotropy Model

A completely different model, the magnetochiral anisotropy (MCA) model, was developed by Wagniere & Meier (1982), explored experimentally by Rikken &

Raupach (2000), and developed further by Barron (2000). In this model, the interaction between photons from an intense source, for example, a supernova, and molecules in a magnetic field, which in principle could also be provided by the supernova, would produce a chirality-dependent destruction effect on the amino acids. This effect does satisfy the conditions of a truly chiral effect (Barron 2000; Avalos et al. 1998; see Appendix B). Although this model suffers from some of the same problems as some other models, most of these can be overcome.

In this particular effect, the dielectric constant of a medium has a term that depends on ***k*** and ***B***, where the vector ***k*** is the wave vector of the incident light (and is in its direction of travel) and ***B*** is an external magnetic field. Because of this term, the dielectric constant increases when incident light travels in the direction of the external magnetic field and decreases when the incident light travels opposite to the direction of the external field. This effect also depends on the enantioselectivity of the material. That is, if the material through which the light passes is optically active (as is the case for amino acids), the effect will have the opposite sign for L- and D-enantiomers. The MCA model is particularly interesting as it does not depend on any polarization of the incident light. The net effect is that one enantiomer absorbs more of the incident light (and thus is preferentially destroyed) than the other, much like in the SNAAP model. In the case of the MCA model, the effect is independent of the polarization of the incident light, which makes it potentially attractive for most astrophysical environments and circumvents one of the potential problems for the CPL model. A schematic diagram of this effect is shown in Figure 9.5 (Wagniere & Meier 1982). Experimental studies on this effect have resulted in *ee*'s on the order of 10^{-4} for chiral molecules (Rikken & Raupach 2000).

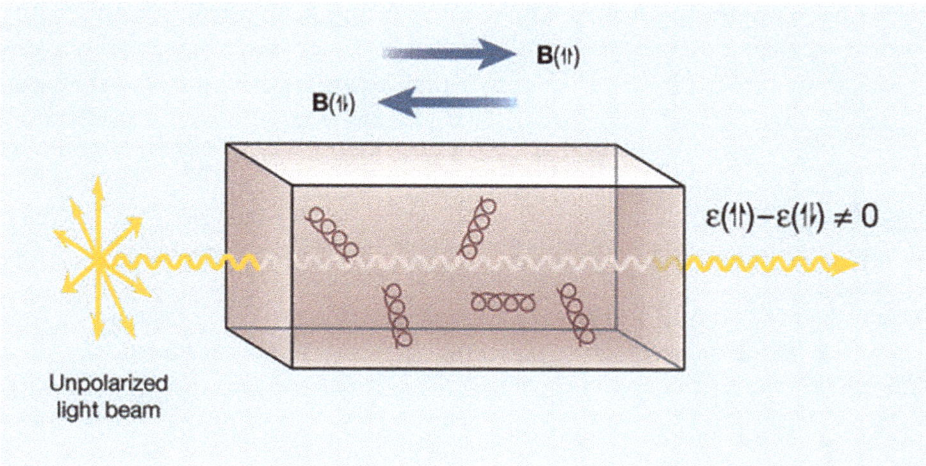

Figure 9.5. An unpolarized light beam passes through a solution of resolved chiral molecules (represented by small helices) in a static magnetic field either parallel B(↑↑) or antiparallel B(↑↓) to the propagation direction. The absorption coefficients ε(↑↑) and ε(↑↓) are slightly different owing to magnetochiral dichroism. Rikken & Raupach (2000) have now exploited this effect to favor the production of one enantiomer. Reprinted by permission from Springer Nature, image courtesy of L.D. Barron (2000).

The issue the MCA model needs to address is, as with the SNAAP model, that a conventional core-collapse supernova cannot provide the photons for the MCA model. The magnetic field drops below the background field after less than 1 au, and that is also the distance that the supernova-to-be extends in its red giant phase, surely cooking any preexisting amino acids that would be within reach of the magnetic field from the nascent neutron star or black hole resulting from the supernova. This problem might be circumvented by a Wolf–Rayet (WR) star, as with the SNAAP model, although the temperature in the cloud surrounding the WR star would again probably destroy any amino acids that were within the magnetic field's processing range. Grains are thought to be created in the cloud of a WR star, but only at distances well beyond the extent of the magnetic field (Barron 2000). Passing meteoroids might be processed by the resulting supernova without complete destruction, so that is one possible solution. However, the passing meteoroids would not be likely to produce a large enough population of meteorites to explain those that have arrived on Earth.

A massive-star–neutron-star binary might work for this model, just as it does for the SNAAP model. One difference between the models is in the processing of meteoroids or planets that are in the accretion disk. As the meteoroids formed, they would be comprised of grains that might be processed by the neutrinos from the neutron star and subsequently agglomerated into meteoroids or planets, in which case they would have processed amino acids throughout. But, as may be more likely, those that formed as meteoroids and then were processed by the neutrinos from the exploding WR star would also have processed amino acids throughout. By contrast, photons from the supernova would only process amino acids on their surfaces. They would not have the same volume of processed amino acids that those from the SNAAP model would have, regardless of the source of the neutrinos. Furthermore, the chirality-selected amino acids would not be likely to survive transport through the atmosphere of a planet on which they might ultimately land.

However, even this might be overcome. The shock wave from the exploding supernova might cause some of the now-processed grains from the accretion disk to form larger objects, and they would have processed amino acids throughout the meteoroid. The larger objects that had been formed in the disk, however, would have had only their surface amino acids processed.

Nonetheless, the MCA model, if coupled to the massive-star–neutron-star binary, does appear to be a viable candidate for the creation of enantiomeric amino acids in outer space.

Might the MCA model and the SNAAP model both have been responsible for the creation of life? Further work, especially more detailed estimates of the *ee*'s produced, will be necessary to determine if this is a possibility.

9.2.5 Chiral Selection via Chemical Selection in Clay

A study by Fraser et al. (2011) found that chiral selection of amino acids can occur in smectite clays such as vermiculite. These clays have expandable interlayer spacings, and the fresh layers exhibit significant chiral enrichment of either left- or right-handed amino acids as they are laid down, depending on the specific amino acid. The clays are certainly environments that could have existed on early Earth, so this immediately becomes a potential way to perform chiral selection.

Of course, it may be problematic for models of this type of chiral selectivity that some amino acids develop left-handed selectivity while others develop right-handed selectivity, unlike the monolithic left-handed chirality found in nature. Furthermore, this model does not create or destroy one chirality, but rather separates them spatially. However, if either environment allowed molecular replication more rapidly than the other, or if the ingredients for autocatalysis were more available at one site than the other, this could explain amino acid chirality selection. It is important to point out the possibility that other substances that might have existed in early Earthly environments could also produce this effect (such effects have been seen in quartz and calcite, as described in detail in Hazen 2005), and it would be interesting to sum over all such substances, if that were possible, to see what the net effect would be. Most specifically, if other substances also selected different chiralities in different amino acids, and in some cases these tended to cancel the chiralities selected in clay, this would be unfavorable to the clay model.

9.3 Amplification via Chemical Catalysis

The issues of both chirality production and amplification were studied in a seminal paper by Gol'danskii & Kuz'min (1989). As noted in Chapter 1, they observe that chiral purity may be the apparatus for self-replication, without which it could not occur. They go on to show that the drive toward homochirality could occur either because one handedness achieved an advantage, for example, one chirality might have become more prevalent than the other because of one of the external forces described above, or because the chemistry that creates new molecules from the required constituents might favor one chirality over the other.

In the latter context, they developed the model for amplification originally suggested by Frank (1953) and advanced by several astrobiologists since (see, for example, Kondepudi & Nelson 1985). As Frank originally conceived it, he hypothesized a substance that could act as a catalyst for its own production and an anticatalyst for the production of its chiral opposite. In this model, the left-handed molecule M_L and the right-handed molecule M_R are both made from constituent molecules A and B, as would be the case with any amino acid. Once made, they can drive autocatalysis, in which they can guide the synthesis of new molecules of their same handedness from additional molecules A and B. Finally, they can also combine to form a new molecule A′, destroying one M_L and one M_D in the process. We have rewritten these three statements as the following equations, where we have indicated to the right of each equation the rate constants, k, that dictate the speed at which each reaction occurs:

$$A + B \rightarrow M_L(k_1{}^L); \qquad A + B \rightarrow M_R(k_1{}^R)$$
$$A + B + M_L \leftrightarrow 2M_L(k_2{}^L); \qquad A + B + M_R \leftrightarrow 2M_R(k_2{}^R)$$
$$M_L + M_R \rightarrow A'(k_S), \qquad .$$

As noted by Gol'danskii & Kuz'min (1989), Frank's original model did not have either of the reactions in the first line, nor did he include the possibility that the second set of reactions could go both ways.

What Gol'danskii and Kuz'min showed is that the relationship between the rates of these reactions can determine a critical value such that, if that value is exceeded, the succeeding reactions will drive the abundances toward a single chirality. With particular values of the reaction rates, the reactions involving molecules of one chirality might be formed over another, that is, one could have asymmetric autocatalysis. They also note that this may be a consequence of the dynamic properties of the system itself and the process by which the mirror isomers (molecules of opposite chirality, but with identical properties in all other respects, e.g., melting and boiling points, solubility, etc.) undergo transformations. Gol'danskii and Kuz'min claimed that these reactions could produce a single chirality even without an initial advantage, that is, just from statistical fluctuations in the densities of the molecules.

However, an asymmetry in the reaction rates could also be important for amplifying existing enantiomerism via asymmetric autocatalysis, wherein the asymmetry favored reactions going toward one chirality over the other. This is an effect that has been demonstrated in a handful of laboratory experiments, although not necessarily for amino acids (Soai et al. 1995; Soai & Sato 2002; Breslow & Levine 2006). More generally, one could enhance the drive toward homochirality by providing any advantage that would enable the dominance of one chirality over the other. Gol'danskii and Kuz'min divide the selective enhancement advantages into two classes, local and global. The local class would include an advantage that would occur only in a specific restricted location. This might include combinations of magnetic and electric fields, or CPL. Global advantages might include the effects of β-decay or some energetic shift between the left- and right-handed molecules, as discussed above, that always produced the same chirality.

9.4 Laboratory Experiments and Theoretical Developments

Several laboratory experiments that show how amino acids might be produced in cosmic dust grains have now been performed. Bernstein et al. (2002) showed that nonchirally selected amino acids (glycine, alanine, serine) could be produced in the laboratory via ultraviolet photolysis of interstellar ice analogs (H_2O, NH_3, CH_3OH, HCN). These would be racemic; chiral selection would have to come later. Studies of this type have been going on for many years; a review article by Allamandola (2008) gives an idea of the scope of this work. Lee et al. (1996) demonstrated that the amino acid glycine could be produced in an environment that simulated ice-coated dust grains in the interstellar medium. This experiment irradiated ice films that contained some of the basic molecules that have been observed in the cosmos, notably ammonia, NH_3, and methane, CH_4, with ultraviolet photons. Of course, the same ultraviolet radiation could destroy the molecules that had been produced. Thus, another important result of the work of Lee et al. (1996) was that it also demonstrated that some of the amino acids so formed would survive ultraviolet radiation at some abundance level, particularly if the molecules existed in a dense part of a molecular cloud. This would provide shielding to some extent from the destruction. In any event, a balance between molecular production and molecular destruction would be established, assuming an equilibrium situation, so that some net amino acids would ultimately be created.

Ultraviolet irradiation is not an essential feature of cosmic organic molecule production. Bennett et al. (2011) showed that energetic electrons, which simulate cosmic rays and can be very abundant in the cosmos, especially in the vicinity of a star (as might exist in solar flares), could produce formic acid, HCOOH, when the electrons hit mixtures of water and carbon monoxide ices.

Woon (2011) investigated theoretically several favorable reaction pathways to produce some interesting organic molecules. He found that the molecules HCOOH, CH_3OH, and CO_2 could be created on simulated icy grain mantles from very low-energy reactions, as would be expected in the cold cosmic environment, when the icy grain mantles interacted with several ions.

In a study of creation and amplification in the interstellar medium, Garrod et al. (2008) developed a model in which chiral replication of complex molecules would occur in the warmed icy outer shells of grains. Chemical replication would be catalyzed by radicals that are combinations of atoms that are not necessarily chemically stable (for example, H, OH, CO, CH_3, NH, and NH_2) and are created by the interactions of high-energy cosmic rays with previously formed molecules. The grains would have to be warmed above the ambient temperature of the molecular clouds to enhance the mobility of the heavy radicals, but presumably this would occur periodically as the grains passed near a star and were subjected to the star's radiation.

9.5 Terrestrial Amplification

The ability of a collection of molecules exhibiting a small enantiomeric excess to amplify dramatically that excess has been demonstrated in several works, even in some environments that might plausibly exist on many planets. We will give brief descriptions of several of them, which avoid a lot of involvement with organic chemistry.

- Soai et al. (1995) began with a left-handed chirality of 2% of 5-pyrimidyl alkanol treated with disopropylzinc and pyrimidine-5-carboxaldehyde. After performing an evaporation, they found that autocatalysis enabled by a chiral catalyst had increased the handedness to 85%.
- Soai & Sato (2002) began with a left-handedness of only 0.05% of methyl mandelate. Again, performing an evaporation, this was found to have been autocatalyzed to a much higher enantiomeric excess. With the same procedure, leucine (an amino acid), initially at a handedness of 2%, was enhanced to >95%.
- Mathew et al. (2004) began with an initial handedness of 5% of proline (an amino acid). After the evaporation, this had a handedness of 65%. But they

observed the interesting feature that the rate at which the reaction occurred accelerated as the chiral fraction increased.

- Finally, Ronald Breslow and his student, Mindy Levine (Breslow & Levine 2006) began with 1% chiral samples of right- or left-handed phenylalanine, which is also an amino acid. The samples were amplified to a chirality of 90% by two evaporations that precipitated out the nonchiral component. The conditions under which these experiments were conducted simulated conditions that could have plausibly existed in Earthly environments.

These experiments are interesting in their own right, but they could have profound implications for the drive to achieve homochirality of the amino acids and of other life-related molecules. These experiments show that any model that can produce some preferred chirality, even at a low level, stands a good chance of triggering a homochiral environment on its chosen planet.

Although the requirement of water in most or all of these experiments would seem to make these mechanisms unlikely in the interstellar medium, dust grains do develop icy surfaces, and a slight warming of these shells from the temperature characteristic of the interstellar medium could provide the conditions in which amplification could occur. Thus, the possibility of amplification in space should not be ignored. However, the water-based experiments are certainly relevant for providing a second stage of amplification once the enantiomeric molecules arrived on Earth. In several billion years of Earthly existence, a lot of meteorites might be expected to have delivered a lot of enantiomeric amino acids to Earth. If most of them had the same chirality, this would surely support the drive toward homochirality.

9.6 Concluding Comments

We previously mentioned the model that we and our colleagues, Kajino, Onaka, and Mo, developed (Boyd et al. 2018; Famiano et al. 2018a, 2018b). This model relies on the weak interaction and on the neutrinos from supernovae. However, in addition to these commonalities with the Cline hypothesis (Cline 2005), our model also requires the strong magnetic field produced by a core-collapse supernova and absolutely requires a non-spin-zero nucleus in order to couple to the molecular chirality. Although this might be satisfied by the hydrogen in the amino acids, it turns out that the effect that hydrogen would produce is not nearly as definitive as that from ^{14}N, a basic constituent of all amino acids. Nitrogen nuclei are thus crucial to our Supernova Neutrino Amino Acid Processing (SNAAP) model.

If any of the suggested scenarios is to succeed, some amplification, although usually a lot of it, is essential. The SNAAP model appears to come close to producing enantiomerism in outer space at the level seen in meteorites, which has not been demonstrated for any other model. For example, in the CPL model, the chirally selected molecules might replicate in outer space. Although this might occur on the warmed surfaces of dust grains that are known to exist in outer space (Soai & Sato 2002), once the chirally selected molecules are bound up in a meteoroid and delivered to Earth in a meteorite, all known models require that they be amplified again.

Amplification once the molecules have arrived on Earth appears to be readily possible. The evaporation experiments discussed above could readily produce the additional amplification that would ultimately produce homochiral molecules. It appears, however, that the SNAAP model presents the most robust scenario for the production of the initial enantiomerism of amino acids at a sufficiently high level that would lead to this ultimate homochirality.

References

Akaboshi, M., Noda, M., Kawai, K., Maki, H. & Kawamoto, K. 1978, OLEB, 9, 181

Akaboshi, M., Noda, M., Kawai, K., Maki, H. & Kawamoto, K. 1982, OLEB, 12, 395

Akaboshi, M., Noda, M., Kawai, K., et al. 1981, OLEB, 11, 23

Allamandola, L. J. 2008, in Chemical Evolution Across Space and Time—From the Big Bang to Prebiotic Chemistry, ed L. Zaikowski, & J.M. Friedrich (Washington, DC: American Chemical Society), 80

Avalos, M., Babiano, R., Cintas, P., et al. 1998, ChRv, 98, 2391

Bailey, J. 2001, OLEB, 31, 167

Bailey, J., Chrysostomou, A., Hough, J. H., et al. 1998, Sci, 281, 672

Barqueno, P. & Perez de Tudela, R. 2007, OLEB, 37, 253

Barron, L. D. 2000, Natur, 405, 895

Bennett, C. H., Hama, T., Kim, Y. A., Kawasaki, M. & Kaiser, R. I. 2011, ApJ, 727, 27

Bernstein, M. P., Dworkin, J. P., Sandford, S. A., Cooper, G. W. & Allamandola, L. J. 2002, Natur, 416, 401

Bonner, W. 1991, OLEB, 21, 59

Bonner, W. 2000, Chirality, 12, 114

Boyd, R. N., Famiano, M. A., Kajino, T., Onaka, T. & Mo, Y. 2018, ApJ, 856, 26

Breslow, R. & Levine, M. S. 2006, PNAS, 103, 12979

Burkov, V. I., Goncharova, L. A., Gusev, G. A., et al. 2008, OLEB, 38, 155

Cline, D. B. 2005, Chirality, 17, S234

Conte, E. 1985, NCimL, 44, 641

deMarcellus, P., Meinert, C., Nuevo, M., et al. 2011, ApJ, 727, L1

Ehrenfreund, P., Bernstein, M. P., Dworkin, J. P., Sandford, S. A. & Allamandola, L. J. 2001, ApJ, 550, L95

Famiano, M. A., Boyd, R. N., Kajino, T. & Onaka, T. 2018a, NatSR, 8, 8833

Famiano, M. A., Boyd, R. N., Onaka, T., Kajino, T. & Mo, Y. 2018b, AsBio, 18, 190

Frank, F. 1953, BBA, 11, 459

Fraser, D. G., Fitz, D., Jakschitz, T. & Rode, B. M. 2011, PCCP, 13, 831

Fukue, T., Tamura, M., Kandori, R., et al. 2010, OLEB, 40, 335

Garrod, R. T., Weaver, S. L. & Herbst, E. 2008, ApJ, 682, 283

Gledhill, T. M. & McCall, A. 2000, MNRAS, 314, 123

Gol'danskii, V. I. & Kuz'min, V. V. 1989, SvPhU, 32, 1

Gusev, G. A., Kobayashi, K., Moiseenko, E. V., et al. 2008, OLEB, 38, 509

Hakala, P. J., Piirola, V., Vilhu, O., Osborne, J. P. & Hannikainen, D. C. 1994, MNRAS, 271, L41

Hazen, R. M. 2005, Genesis: The Scientific Quest for Life's Origins (Washington, DC: Joseph Henry Press)
Kondepudi, D. K. & Nelson, G. W. 1985, Natur, 314, 438
Lee, D. H., Granja, J. R., Martinez, J. A., Severin, K. & Ghadiri, M. R. 1996, Natur, 382, 525
Lee, T. D. & Yang, C. N. 1956, PhRv, 104, 254
MacDermott, A. J., Fu, T., Nakatsuka, R., Coleman, A. P. & Hyde, G. O. 2009, OLEB, 39, 459
Mann, A. K. & Primikov, H. 1981, OLEB, 11, 255
Mason, S. F. 1984, Natur, 311, 19
Mason, S. F. & Tranter, G. E. 1983, JCS ChCom, 117
Mason, S. F. & Tranter, G. E. 1984, MolPh, 53, 1091
Mason, S. F. & Tranter, G. E. 1985, RSPSA, 397, 45
Mathew, S. P., Iwamura, H. & Blackmond, D. G. 2004, Angew. Chem., 43, 3317
Meinert, C., Bredehoft, J. H., Filippi, J.-J., et al. 2012, Angew. Chem., 51, 4484
Miller, S. L. 1953, Sci, 117, 528
Miller, S. L. & Urey, H. C. 1959, Sci, 130, 245
Modica, P., Meinert, C., de Marcellus, P., et al. 2014, ApJ, 788, 79
O'Leary, M. 2008, Anaxagoras and the Origin of Panspermia Theory (New York: iUniverse)
Rikken, G. I. J. A. & Raupach, E. 2000, Natur, 405, 932
Rode, B. M., Fitz, D. & Jakschitz, T. 2007, Chem. Biodiversity, 4, 2674
Soai, K. & Sato, I. 2002, Chirality, 14, 548
Soai, K., Shibata, T., Morioka, H. & Choji, K. 1995, Natur, 378, 767
Takano, Y., Takahashi, J. I., Kaneko, T., Marumo, K. & Kobayashi, K. 2007, E&PSL, 254, 106
Tranter, G. E. 1985a, CPL, 120, 93
Tranter, G. E. 1985b, Natur, 318, 172
Tranter, G. E. 1985c, CPL, 115, 286
Tranter, G. E. 1986, J. Theor. Biol., 119, 467
Tranter, G. E. 1987, CPL, 135, 279
Ulbright, T. L. V. & Vester, F. 1962, Tetrahedron, 18, 629
Vester, F., Ulbright, T. L. V. & Krauch, H. 1959, NW, 46, 68
Wachterhauser, G. 1988, MicRv, 52, 452
Wachterhauser, G. 1992, Prog. Biophys. Mol. Biol., 58, 85
Wagniere, G. & Meier, A. 1982, CPL, 93, 78
Woon, D. E. 2011, ApJ, 728, 44
Wu, C. A., Ambler, E., Hayward, R. W., Hoppes, D. D. & Hudson, R. P. 1957, PhRv, 105, 1413

Richard N Boyd and Michael A Famiano

Chapter 10

Beyond the Amino Acids

10.1 How are More Complicated Molecules Made?

Of course, just making amino acids does not create life. The amino acids somehow must manage to get together to form very complex molecules, the proteins. But one protein may have hundreds to thousands of amino acids, and the 20 amino acids that are relevant to Earthly life could form a huge number of different proteins, only a small fraction of which would end up being useful to maintain life. So we are left to wonder how the amino acids get their instructions to form only the useful proteins, or whether others are made but subsequently wither away.

Although there are a lot of questions associated with the origin, evolution, and operation of cells, this one does have an answer, courtesy of our DNA (deoxyribonucleic acid). DNA contains the instructions for making the essential proteins, down to the specific amino acid structure. DNA is constructed from four nucleobases, adenine, guanine, cytosine, and thymine, abbreviated A, G, C, and T. We will also encounter another nucleobase, uracil, denoted by U, which is similar to T, but which only occurs in RNA. We will come back to the nucleobases in a bit; their specific chemical character is crucial for replication, one of the basic criteria for achieving the status of a living being.

Although DNA has the instructions for forming the essential proteins, it must communicate these instructions efficiently so that the correct proteins get made, and that they get made correctly. This is done via RNA, ribonucleic acid. RNA is made of AGCU and comes in several varieties. Messenger RNA, or mRNA, reads the protein-making instructions from the DNA and takes that information to the ribosomes, the places within the cell where the proteins are made. These have a "slot" into which the mRNA feeds the instructions, selecting the amino acids and assembling them segment by segment until the protein is complete.

While the assembly is taking place, somehow the right amino acid has to be identified and brought to the growing protein chain at the right time. This function is performed by the transfer RNA, tRNA. The mRNA tells the tRNA which amino

doi:10.1088/978-0-7503-1993-5ch10

acid it needs next, and the tRNA finds it, attaches the amino acid to itself, and delivers it to the correct site on the growing protein. This process continues until the protein is formed, at which time the mRNA sends a signal to the ribosome, indicating that the process is complete and the newly created protein can be sent on its way to perform its designated task.

Of course, DNA does more than just provide the instructions for protein assembly. Davies (1999) has given a nice description, which will form the basis and level for what follows.

In the 1950s, Francis Crick and James Watson discovered that DNA has a double-helix structure (Watson 2001). The double strands are attached by many cross-links, which is where the nucleobases come in, and are paired in such a way that A is always linked with T and C with G. Thus, the two strands are complementary, and when it comes time for DNA to replicate, the two strands unwind, and each strand will combine with their new bases, maintaining their nucleobase pairings, to form a new strand that is identical to the one from which it just unwound. Figure 10.1 shows a section of a DNA double helix.

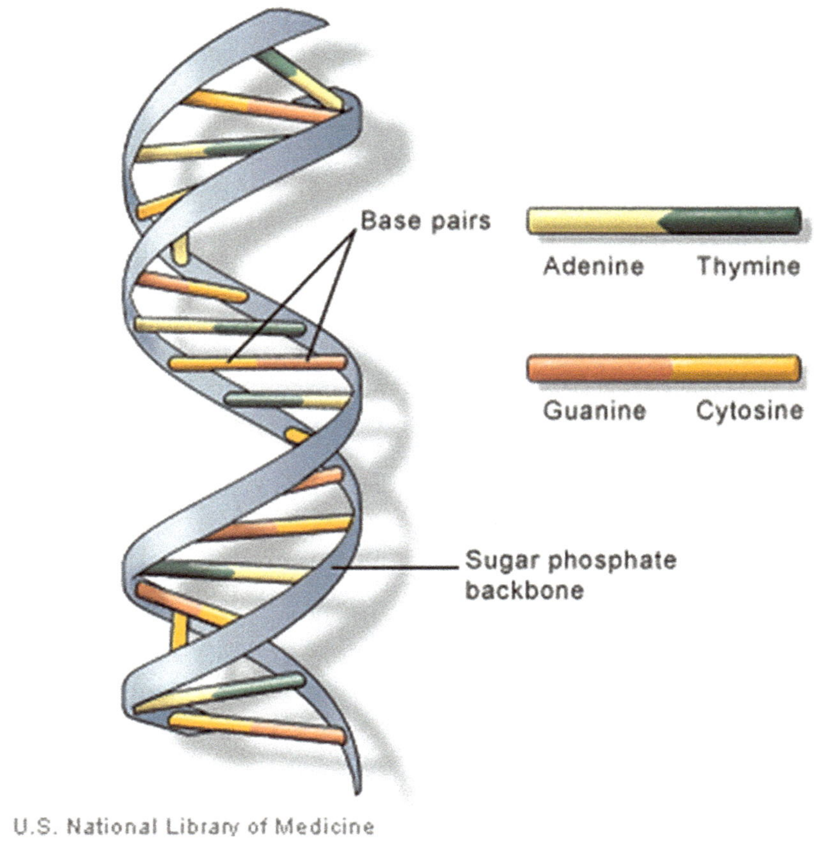

Figure 10.1. A section of a DNA double helix, showing the A–T and C–G pairing between the different strands of the double helix. Copyright US National Library of Medicine.

One of the basic requirements for life is the ability to reproduce, and this clearly must persist for generations. It is this complementarity of DNA chains that produces its required unwavering replication. Human beings have been around for many generations and that serves as testimony to the accuracy of the replication process. Variations do occur, but these are relatively rare, given the incredible number of atoms in DNA. Furthermore, DNA has mechanisms for correcting errors of replication. Of course, occasionally an error will remain uncorrected; this will result in a mutation. Although most mutations are benign or destructive, once in a while an advantageous mutation occurs, and this will provide its lucky recipient with a survival advantage, giving the entity with the new DNA, and whatever progeny it may subsequently conceive, a potential evolutionary gain.

10.2 How were the More Complex Molecules Created?

The description of how the proteins are made and how DNA replicates itself, of course, does not describe how everything got started. We do know that amino acids are produced in outer space even with a start of the correct chirality. This comes from the space explorers *Stardust*, *Hayabusa*, and *Rosetta* and especially from the meteoritic evidence. But how did nature get from the amino acids to the more complex molecules, the proteins and DNA? One model suggests that peptides (formed from the amino acids) came first, and from these the more complex molecules of life, DNA and RNA, were formed; this is known as the protein world. Peptides contain at least two, but not more than 20, amino acids. The more complex molecules would then have formed initially as oligomers, single-stranded nucleic acid fragments. The other model suggests that those more complex molecules had to be formed first, because without them how would the proteins get the instructions they require to form? This is called the RNA world, which was originally proposed by Gilbert (1986). But this model has to deal with a difficult question: where did the RNA strand come from?

However, there is now experimental evidence that amino acids can form peptides, and once that huge step has been taken, formation of more complex molecules can proceed. Surely the first ones were simple, but as mutations occurred and natural selection oversaw the triumph of the more dominant forms, eventually, after many more giant evolutionary steps, the complexity that we now know was achieved. There are still many missing or poorly understood steps, but the model at least seems plausible.

It has been suggested that peptides form through salt-induced peptide-formation reactions, or SIPF reactions (Fitz et al. 2007). Although based on theory, the evidence for the reactions by which the peptides would be synthesized in a high-concentration salt-water environment is sufficiently well developed (Tauler & Rode 1990; Rode & Schwendinger 1990) to warrant a detailed description of the process. The critical reactions were apparently catalyzed by copper ions in the salty environment (Fitz et al. 2007). The copper served (Rode & Schwendinger 1990; Rode et al. 2007) to bring the amino acids together into an appropriate position and polarization state to encourage peptide formation to take place. Also essential to the process would be the dehydrating agent NaCl (Fitz et al. 2007; Rode & Schwendinger 1990). Assuming a reaction set that is catalyzed by glycine (indicated as Gly) and X being another amino acid, the reactions are as follows. Dashes indicate bonds, and the process indicated by these equations forms a new chemical of the two X amino acids (Rode & Schwendinger 1990):

$$\mathrm{Gly} + \mathrm{X} \rightarrow \mathrm{Gly} - \mathrm{X}$$
$$\mathrm{Gly} - \mathrm{X} + \mathrm{X} \rightarrow \mathrm{Gly} - \mathrm{X} - \mathrm{X}$$
$$\mathrm{Gly} - \mathrm{X} - \mathrm{X} \rightarrow \mathrm{Gly} + \mathrm{X} - \mathrm{X},$$

where the dashes indicate bonds, i.e., X – X is a molecule consisting of two of the X amino acids.

The review article by Rode, Fitz, and Jakschitz (Rode et al. 2007) deals with this question, and we summarize their conclusions.

How was life initiated? Could amino acids and peptides have provided the necessary conditions needed to define life processes. Questions of self-replication and of autocatalytic processes have to be dealt with. These certainly occur with RNA and DNA, but are these types of processes also possible with peptides alone? It has been demonstrated that some peptides can reproduce themselves from smaller constituents (Isaac & Chmielewski 2002; Lee et al. 1996; Yao et al. 1998), thus replicating their structure and their properties. But it has also been found that amino acids can catalyze peptide-formation processes, an example of autocatalysis in the production of oligomers (Suwannochot & Rode 1999). Thus, chemical evolution towards primitive life forms may well have been initiated with a number of amino acids, forming peptides and, later on, the proteins. These pieces seem to argue in favor of a protein-first world, starting from smaller peptides and forming in the end some cell-like organisms with the first simple characteristics of life. Of course, one would not expect to find these forms in existence now, or even in the fossil record. The more evolved forms would certainly have consumed them.

However, an interesting twist occurred, due in part to Stanley Miller. The original Miller–Urey experiment showed that at least some amino acids could be formed in the conditions and within the environments thought to characterize early Earth. However, recent experiments have shown that a completely different environment is also capable of producing amino acids and some nucleobases as well. The first such experiment, performed by Miller, consisted of a solution he had prepared 25 years before he analyzed it. The experiment consisted of a dilute solution of ammonium cyanide, NH_4CN, that was kept at two temperatures, −20 °C and −78 °C, for 25 years. Although chemical reactions are generally thought to slow down as the temperature decreases, the rates at which the reactions occur also depend on the densities of the entities that need to come together in order to react. As ice crystals form, they tend to drive out any impurities, which then collect at the interfaces between the ice crystals, thereby vastly increasing the densities of the impurities. This is known as eutectic freezing. This result was reported by Fox (2008), who noted "Chemically speaking, [eutectic freezing] transforms a tepid seventh-grade school dance into a raging molecular mosh pit." The results were published in 2000. Levy et al. (2000) found a simple set of amino acids, as well as adenine and guanine, two of the nucleobases. Miyakawa et al. (2002) also published results of eutectic freezing experiments.

Lower temperatures have additional advantages. Cyanide evaporates more readily than water, so operating at elevated temperatures tends to eliminate that constituent from a liquid mix. In addition, the nucleobases are rather fragile molecules, so lowering the temperature helps preserve them, extending their lifetime (Fox 2008) from days to centuries.

Additional experiments confirmed the result of Levy et al. (2000). Trinks et al. (2005) conducted an experiment in which they seeded artificial sea water that was spiked with the RNA nucleobases adenine, cytosine, and guanine, and with an RNA template, a single-strand chain of RNA, to guide the formation of new strands of RNA. Then they froze the mixture for a year. New strands of RNA assembled precisely as instructed by the existing one. What Trinks et al. (2005) found was that long strands of RNA molecules—up to 400 nucleobases long—had been formed. Although this experiment does not show how the RNA could have been formed initially, it does show that once such a strand existed, further growth of chains would occur.

These results certainly give hope to finding life on places like Europa (a moon of Jupiter), which is generally thought to be covered by a several-kilometer-thick layer of ice. However, it might also support the idea that it is not only amino acids, but also nucleobases and even strands of RNA that might be formed in the cold confines of outer space. So perhaps the RNA world is not dead. But the question still arises as to where the template strand of RNA came from.

Of course, just having homochiral amino acids and nucleobases does not get us to the point where these chemicals evolve into living beings. Scientists certainly do not yet understand the evolutionary processes through which this happens. However, a viable site where the myriad chemical reactions required for this to happen from the molecules of life that are delivered to Earth from outer space has been suggested (Brasier et al. 2011): pumice. It is very porous, which gives it an extremely high surface-to-volume ratio, and presents large surface areas on which the constituents necessary to create complex molecules and replicate the basic molecules of life, the amino acids and the nucleobases, could congregate. Pumice also has a very low density, which makes it capable of floating on the surface of water, at least until it saturates. This allows it to pile up on shorelines so as to make large chunks of this chemical cauldron. Furthermore, the chunks of pumice might continue to serve their catalytic function even when they sink, especially if they happen to end up near hot ocean vents, to which we will return in a subsequent section.

However the molecules of life formed, and however life evolved from them, somehow life *did* get started from the less complex molecules that preceded it. Might there have been just a single life form that evolved in a variety of different ways to form the plethora of living things that now inhabit Earth and that have inhabited it over the eons it has existed? The model that biologists have posited for this traces back to the last universal common ancestor, LUCA (see Plaxco & Gross 2006 for a more extensive discussion). What biologists have found is that all living creatures have many biological features in common. For example, even though no actual

LUCA has ever been unearthed by paleobiologists, the commonality of the species that "she" originated suggest that she relied on her DNA to store her genetic information, and that she used the same 20 (presumably left-handed!) amino acids to manufacture a couple hundred proteins that her cells used to perform their various functions. Why has her fossil remnant never been found? She was most likely a very primitive creature, perhaps even a single cell, so she might not be something that would be readily recognizable after having been a fossil for a few billion years. And LUCA may not have been unique; other LUCAs may have been created at about the same time she was, and even in several different locations around our planet, but she, or several others like her, was just the most successful in the competition for survival that had to have ensued.

10.3 Extremophiles on Earth

Of course, it is possible that very different life forms evolved from the same basic chemicals we are made of. Indeed, it may be difficult to even recognize the critters that inhabit other planets, or even obscure places on Earth. This gets us back to our discussion of the definition of life. Although we may not want to wait around for our candidate life forms to multiply, it would certainly be essential for them to metabolize their food, whatever form that might take. That might be the most plausible means of detecting that our sample contains something that is alive. But even this might take on forms with which we are not familiar; even on Earth, critters have been found that eat things that would literally poison the life forms we know and love. For example, extremophiles that happily consume sulfur have been found (Norlund et al. 2009). *Ferroplasma acidarmanus* has been found to grow at a pH of 0.0 in highly acidic mine drainage (Edwards et al. 2000) from Iron Mountain in California. There, it apparently thrives on a mixture of sulfuric acid that is laced with high levels of copper, arsenic, cadmium, and zinc. In the case of sulfur-eating bacteria, these were suggested as being capable of cleaning up the sulfuric acid mess that results from mine tailings. Thus, some of our more extreme "relatives" with which nature has presented us may be better at maintaining our natural resources than the intelligent life on this planet of which we like to think we are the ultimate form. In any event, the search for life must cast a wide net for possibilities that might not look very much like the living things we could identify easily.

However, we know of living things that are more than just those with dietary oddities. The so-called extremophiles are known to push the limits of temperature, acidity, drought, and probably a lot of other things well beyond what humans could tolerate. To quote Boston (2011, p. 3), "When we look at our own planet's most challenging environments, we are really looking for clues to what may be the normal conditions on other planets. We want a hint of what we may be searching for when we investigate those other worlds for signs of life." So, our searches of our Earthly environment for the weirdest critters may provide helpful signals for our searches for life beyond Earth. Boston's subtitle for her article was "What is extreme here may be just business-as-usual elsewhere."

Let us look at a few examples of extremophiles in enough detail to give the reader some idea of what extreme really means. There are hydrothermal vents in the deep sea that spew forth material: these are called either white smokers or black smokers. The white smoker at the Champagne vent, northwest Eifuku volcano, Marianas Trench Marine National Monument, produces liquid carbon dioxide. Recall that in the deep ocean the pressure can be several hundred times higher than it is at Earth's surface; this is the reason why carbon dioxide can even exist as a liquid (it is either a gas or a solid at atmospheric pressure). Black smokers are little ecosystems in themselves. They spew forth a variety of chemicals that build up along their sides to produce a sort of chimney. Although the temperature of sea water near these black smokers is about 2 °C, just above the freezing point of water at Earth's surface, the temperature of the material coming from the vent can be as high as several hundred degrees centigrade, or several times the boiling temperature of water at Earth's surface. Of course, water does not boil at that ocean depth, again because of the high pressure.

Despite their temperature extremes, black smokers have been found to be home to a variety of living entities. In particular, Kashefi & Loveley (2003) analyzed microorganisms from the black smoker Finn, located in the Mothra hydrothermal vent field (47°55.46′N and 129°06.51′W) along the Endeavor segment of the Juan de Fuca Ridge in the northeast Pacific Ocean. One of the organisms they studied, strain 121, was isolated at 100 °C, the boiling point of water at atmospheric pressure. Some iron oxide was included in the mixture. Cell growth and iron reduction were monitored. Strain 121 is typical of archaea, which are single-celled organisms, whose cells do not contain nuclei (see the discussion in Chapter 1). It grew at temperatures between 85 °C and 121 °C, and continued to exist even at those extreme temperatures. Because of the ability of this and other hyperthermophiles (defined as organisms that can live at temperatures in excess of 80 °C) to withstand high temperatures, they have been suggested as possible early life forms for Earth, which also existed at an extremely high temperature during the stage when life began (http://www.theguardians.com/Microbiology/gm_mbm04.htm). This point is supported by Davies (1999), who strongly advocates that the microbial life that exists near the black smokers is possibly closely related to the first life forms that existed on Earth.

Heat-loving critters were actually first discovered in the geothermal ponds in Yellowstone National Park, although the Yellowstone extremophiles are not quite as extreme as those from the black smokers. This is described nicely by Rothschild & Mancinelli (2001) and by Bordenstein (http://serc.carleton.edu/microbelife/topics/octopusspring/index.html) for one of Yellowstone's hot springs, Octopus Spring, which is highly alkaline (pH of 8.8–8.3). It is located about eight miles north of the Old Faithful geyser. As described by Bordenstein (http://serc.carleton.edu/microbelife/topics/octopusspring/index.html), the spring has a primary source from which the water flows at 95 °C. Its drainage channels radiate like the arms of an octopus (requiring some imagination), hence the name of the spring.

As the water flows to one part of the accompanying pond, its temperature cools to about 83 °C, but it then increases again to 88 °C when a new surge of hot water is emitted. In this environment, pink filamentous communities thrive, at a distance of about 2 m from the crystal blue pool. Elsewhere in the pond, the water cools to less than 75 °C, a temperature that permits the growth of microbial mat communities that include cyanobacteria, green sulfur bacteria, and green nonsulfur bacteria. One can see the growth of a thermophilic cyanobacterium in this part of the pond. At 65 °C, a more complex microbial mat forms one bacterium on the top overlying other bacteria, including species of a photosynthetic bacterium. Some superb photographs have been taken of Octopus Spring; one of these is shown in Figure 10.2. A picture of a microbial mat is shown in Figure 10.3.

At the other temperature extreme are the cold-loving entities, the cryophiles. These include cyanobacteria, bacteria, fungi, spores, etc. (https://www.livescience.com/38652-what-is-lake-vostok.html). These were discovered in Lake Vostok, in Antarctica, by Richard Hoover of NASA's Marshall Space Sciences Laboratory and S. S. Abyzov of the Russian Academy of Sciences. The organisms they found are thought to have been trapped in the ice about 400,000 years ago. They have existed in a dormant state ever since, but are still metabolizing at an extremely low level. Their existence might bode well for finding life forms on cold planets or moons of planets, for example, on Europa, one of Jupiter's moons.

However, the most significant aspect of life existing at cryogenic temperatures may be, as noted earlier, that the amino acids and nucleobases can be assembled via eutectic freezing, even at temperatures so low that ordinary life, at least human life, could not be sustained. However, if life forms tried to exist at such temperatures,

Figure 10.2. The geothermal pond Octopus Spring, in Yellowstone National Park. From NASA Astrobiology Institute. Credit: Amaya Garcia Costas.

Figure 10.3. A microbial mat from one of the Yellowstone National Park thermal ponds. Note the striations showing the different biota that have inhabited this mat during its life. Figure courtesy of David Ward, Ward Lab, Montana State University.

they would need to be very careful; when water freezes inside a cell, it almost always ruptures the cell when it expands from its liquid state to its ice state. But the molecules of life seem to have no difficulty assembling at such temperatures; indeed, the assembly of at least some of those molecules appears to require those temperatures.

Of course, this may have important ramifications on our conclusions about the source of the molecules of life. It might be difficult to imagine cold conditions on the early Earth, although the day–night cycles might have been helpful in that regard, and such cycles might even be important in producing some of the molecules of life. However, low temperatures occur naturally for meteoroids and comets for much of their lives, so it would not be surprising if some low-temperature processes were important in creating the amino acids. As noted in Chapter 6, the molecules produced in the insides of large solid objects included therein would not be subject to processing by ultraviolet radiation. However, they would be sensitive to strong external magnetic fields. They would also be sensitive to neutrino processing! This independence of the size of the objects being processed is an important aspect of the SNAAP model (Boyd et al. 2010, 2011, 2018; Famiano et al. 2014, 2018) that makes it unique among the models devised to explain chirality.

As another example of an extremophile, the bacterium *Deinococcus radiodurans* has been found to be able to withstand both gamma radiation and ultraviolet radiation in extremely high doses. Its radiation resistance apparently revolves around its ability to repair the damage to its DNA that results from the high radiation doses. It does this by reassembling the fragmented DNA (Battista 1997, 1998). Would that humans had this capability!

Add to these the critters that prefer very low or very high pH levels, that is, very acidic or basic environments, those that exist in very salty or very metal-rich environments, those that exist at very high pressure, and others (http://en.wikipedia.org/wiki/Extremophile), and it becomes clear that the possibilities for life are enormous and may lie far from what humans would regard as "normal."

The obvious question to ask, though, is how do extremophiles manage to exist in such extreme surroundings? Rothschild & Mancinelli (2001) deal with this question for several extremophiles. The thing that seems to work best is for them to simply keep the hostile components of the external environment out. For example, *Cyanidium caldarium* and *Dunaliella acidophila* are found at an extremely acidic pH of 0.5. But Pick (1999) and Beardall & Entwisle (1984) found them to have a nearly neutral cytoplasm (the thick liquid that holds all of the internal entities of a cell except the nucleus). This can only be the case if the internal parts of their cells are buffered against the highly acidic external environment. Another possibility is to remove the extreme condition as rapidly as possible. An example is heavy-metal-resistant bacteria. Niles (2000) found that they use an efflux pump to remove zinc, copper, and cobalt, but not mercury, which is volatilized.

If it is impossible to keep the environment out, then extremophiles appear to adopt protective mechanisms, altering their physiology or having specific repair mechanisms. Of particular note in this regard are the nucleic acids, for which function and structure are closely linked. DNA is especially vulnerable to high temperature, radiation, oxidative damage, and dehydration. In some cases, extremophiles have adopted multiple ways to solve the problem of living in a particularly hostile environment.

But it would not be appropriate to leave the extraordinary without mentioning the striking features of the very ordinary, a pond-dwelling paramecium. *Paramecium bursaria* appears green under a microscope because each cell hosts hundreds of chlorella algae that supply it with sugar and oxygen. It can reproduce asexually, splitting into two identical daughter cells, but occasionally docks with another cell to exchange small capsules that hold DNA in order to correct damaged DNA. They can swim 10 times the length of their body in one second. Furthermore, they have almost 40,000 genes, about double the number in a human cell (Grunbaum 2011). Apparently, all of these genes give *P. bursaria* the capability to deal with its necessities. Even ordinary can be pretty extraordinary sometimes!

10.4 And from Outer Space?

Of course, identifying extremophiles and associating their environmental preferences with the conditions on early Earth certainly does not guarantee that they have anything to do with early life. This identification would require that we could reconstruct all intermediate stages of life, a very tall order at present and possibly for a long time to come. Most discouragingly, the best we could ever do would be to reconstruct a plausible succession of living forms, but even that would not prove that these were the actual forms that existed.

In recognition of this, and in an attempt to circumvent the difficulty of identifying successive steps, a group set out in 1960 to simply detect the presence of extraterrestrial beings. This effort was called the Search for Extraterrestrial Intelligence, or SETI, and its founding father was Frank Drake (of the Drake equation, discussed in Chapter 4). Later, Carl Sagan played an important role, both scientific and political, in furthering the SETI program. The main question they had to address was: how do you "look" for extraterrestrial beings? You cannot just go somewhere beyond Earth and look for them (with the possible exceptions of our Moon and Mars) given the difficulties of space travel, the large distances involved, and the limitations of special relativity; you need to try to identify plausible places for life and search there. But space is a big place, so this is an overwhelming challenge. Instead, one has to figure out what would constitute an identification of an extraterrestrial life signal that would be convincing even in the absence of an actual sighting.

What was settled on was a detection of a radio signal from outer space that was at some frequency that would be of particular significance to life. It should be noted that the use of radio signals for detecting the presence of extraterrestrial life had already been suggested just before the beginning of the 20th century by Nicola Tesla, and later by such notables as Guglielmo Marconi and Lord Kelvin. A paper published in 1959 by Cocconi and Morrison suggested searching in the microwave part of the electromagnetic spectrum for life signals, setting the stage for the Drake experiment in 1960. Although there were many frequencies suggested as life signals, technology caught up with politics (usually it is the other way around, if it happens at all) when it became possible for radio telescopes to monitor a huge number of radio frequencies simultaneously. This obviated the need to have the life signal frequency match that of the civilizations for which we were searching.

The advantage of using radio signals to search for life signals is fairly obvious to astronomers: radio telescopes (see Chapter 4) can be huge, can be made of multiple dishes so as to present a huge photon-collecting area, and can have good capability for localizing the source of the signals when the telescopes are spread over the entire diameter of Earth. Radio signals also are not absorbed by Earth's atmosphere, so Earth-based observations are adequate for the task, and large space-borne telescopes are unnecessary.

Unfortunately (or, perhaps, fortunately), no convincing signals have yet been detected. This does not mean that there are no intelligent (by our standards, anyway) beings out there. They might be too far away for us to observe their life signals, they may not have yet achieved the capability to send strong radio signals into space, they may have advanced in their communications beyond the use of radio signals, or they may not be especially interested in trying to connect with what they would identify as less advanced aliens—that would be us. Those are all factors that we discussed in the context of the Drake equation. Searching for extraterrestrials can be tough business! And, in the always challenging fiscal times, even though SETI has been mostly privately funded, obtaining funding to continue such efforts can also be difficult.

In any event, the lack of convincing positive results in searches for extraterrestrials led to what is known as the Fermi paradox, suggested by Enrico Fermi in the 1950s (http://en.wikipedia.org/wiki/SETI). It is stated in a variety of ways, but is summarized by the question, if the universe is teeming with life, as many would suggest, where is everyone? The question can be condensed to three issues: (1) perhaps the assertion that the universe is teeming with (technologically advanced, avoiding the use of the term intelligent) life is wrong, (2) we have not looked sufficiently exhaustively yet, or (3) we have not used a detection technique that would have found them. This triad of possible conclusions would seem to cover all possibilities!

Two extreme answers seem to exist. One asserts that life in the universe is sufficiently rare that there are not that many aliens to detect. The other posits that they are already here, and we just do not know how to recognize them.

So perhaps we have been looking in the wrong places or with the wrong tools for extraterrestrial life. With the advent of the results from the *Kepler* mission (https://www.nasa.gov/kepler/discoveries), more and more candidate planets are becoming known. Surely some of those will be Goldilocks planets, that is, planets that are not too hot and not too cold. And, with modern technology, scientists can search for different signatures of life other than intercepted radio signals, such as the production of methane and other gases that might be a byproduct of life and that could be detected in the planetary atmospheres. Indeed, such searches are moving to the fore (Bhattacharjee 2011).

Then, there is the possibility that Francis Crick might be partially correct in asserting that life was brought to Earth by an intelligent civilization from another planet—the Panspermia hypothesis. As noted by Davies (2010), the visitors might even have left their mark in a way that we would find difficult to detect with our present technology. For example, they might have implanted some specific DNA modification on whatever form of DNA they found when they visited. We must take into account the very real possibility (probability?) that there exist technically more advanced civilizations in our Galaxy than the one with which we are familiar!

It seems obvious that we should seek to communicate with extraterrestrial beings, simply on the basis that more information is better than less information, and besides, the existence thereof is an extremely intellectually interesting question. However, there have been voices raised in opposition to such communications. Perhaps the most notable is that from the late Stephen Hawking, who has observed (http://news.bbc.co.uk/2/hi/uk_news/8642558.stm) that the mathematics involved seem to make the existence of aliens a real possibility. However, he added his voice to those of others who noted that it might be a problem to know for certain how aliens might be identified. He went on to warn against contacting aliens and doing our best to avoid them. In this regard, he observed (http://news.bbc.co.uk/2/hi/uk_news/8642558.stm) that an alien visit to Earth might have similarities to the situation that existed when the Europeans came to America, and how that visitation did not go so well for the natives. Perhaps it would be advisable to keep a low cosmic profile!

Let us give a bit more thought to Davies' suggestion of the possibility of an alien DNA modification. As Davies (2010) and many others note, there is a lot of our DNA that does not seem to be involved in any essential aspect of our lives—it is generally referred to as junk DNA. Might our hypothetical advanced aliens have figured out that they could store a "Hi Earthling" message in our DNA that would not affect the function of the DNA? Perhaps someday we will have become clever enough to read the message. Perhaps they were more insidious and reprogrammed our DNA so that when they return to Earth, they will flip a genetic switch and we will suddenly all change to a slave mentality. The possibilities are limitless.

Nonetheless, the basic chemicals of life are made in outer space. The meteoritic evidence is strong that the basic molecules of life—both the amino acids and the nucleobases—are made far from Earth. Furthermore, the fact that amino acid chirality is established in space would suggest that the molecules that are produced in space are the triggers of terrestrial life. Of course, other planets could also not have avoided being hit by many such meteorites, so all planets that harbor life would have been seeded with these molecules.

10.5 So Are We Alone in the Universe?

As discussed in Chapter 4, perhaps other planetary systems also contain Earth-like planets at the right distance from their stars to allow them to have liquid water, and perhaps even Jupiters to sweep out the space detritus inside their orbits, and thus to protect the inner planets. There are, after all, more than 100 billion stars in our Galaxy alone, and there should be a lot more than one of them that would satisfy the special conditions that Earth enjoys.

But this assumes only that the right physical conditions are necessary for life to begin on a planet. The SNAAP model, and most other models, suggests that something else can influence the possibility of life developing, that being the insertion of chiral amino acids into the planetary system. The models that have been suggested, if they are capable of creating chiral amino acids at all, are probably capable of producing the required effects only in localized situations. It is unlikely that the SNAAP model or any of the other scenarios we discuss in Chapter 8 could populate more than a fraction of the potentially life-harboring planets in the Galaxy with enantiomeric molecules.

So, are we alone in our Galaxy? The SNAAP model and the magnetochiraloptical model are both localized to binary systems having one massive star that could ultimately explode as a Wolf–Rayet star. These are not frequent; they could populate only a small fraction of the Galaxy with chiral amino acids, but it appears quite possible that every one of them that ever existed in our Galaxy would produce the conditions that could eventually produce life if the planets they created had the right physical conditions. Furthermore, they would most likely evolve to become a two-neutron-star merger, so their amino acid production would be enhanced considerably from the binary system yield. The *Kepler* results suggest that a fairly large fraction of the stellar systems having stars of reasonable mass do have planets with the appropriate conditions for life. However, unless someone comes up with a model that produces chiral amino acids over much of our Galaxy, most of the Goldilocks planets being found by *Kepler* are sterile, at least as far as life is concerned.

Finally, the recently published observations that our solar system was produced by the explosion of a Wolf–Rayet star (Dwarkadas et al. 2017) lend strong support to this conclusion. But is there any other system that could provide the essential properties for producing enantiomeric amino acids? Is our system unique? We do not believe that our solar system is unique, but given the somewhat heroic measures that all known models of chiral amino acid production must go through to succeed, systems like ours will not be frequent.

If that is correct, we are probably not alone in the universe, or even in the Galaxy. But our nearest neighbors may be a very long way away.

References

Battista, J. R. 1997, Annu. Rev. Microbiol., 51, 203

Battista, J. R. 1998, in DNA Damage and Repair, Vol I: DNA Repair in Prokaryotes and Lower Eukaryotes, ed J. A. Nickoloff, & M. F. Hoekstra (Totowa, NJ: Humana), 287

Beardall, J. & Entwisle, L. 1984, Phycologia, 23, 397

Bhattacharjee, Y. 2011, Sci, 333, 930

Boston, P.J. 1999, Ad Astra Magazine, 11,

Boyd, R. N., Famiano, M. A., Onaka, T. & Kajino, T. 2018, ApJ, 856, 26

Boyd, R. N., Kajino, T. & Onaka, T. 2010, AsBio, 10, 561

Boyd, R. N., Kajino, T. & Onaka, T. 2011, Int. J. Mol. Sci., 12, 3432

Brasier, M. D., Matthewman, R., McMahon, S. & Wacey, D. 2011, AsBio, 11, 725

Davies, P. 1999, The 5th Miracle: The Search for the Origin and the Meaning of Life (New york: Simon and Schuster)

Davies, P. 2010, The Eerie Silence: Renewing Our Search for Alien Intelligence (New York: Houghton, Miflin, Harcourt)

Dwarkadas, V. V., Dauphas, N., Meyer, B., Boyajian, P. & Bojazi, M. 2017, ApJ, 851, 147

Edwards, K. J., Bond, P. L., Gihring, T. M. & Banfield, J. F. 2000, Sci, 287, 1796

Famiano, M. A., Boyd, R. N., Kajino, T., et al. 2014, Symmtery, 6, 909

Famiano, M. A., Boyd, R. N., Kajino, T. & Onaka, T. 2018, AsBio, 18, 190

Fitz, D., Reiner, H., Plankensteiner, K. & Rode, B. M. 2007, CCB, 1, 000

Fox, D. 2008, Discover 29, 2

Gilbert, W. 1986, Natur, 319, 618

Grunbaum, M. 2011, Disc, May 16

Isaac, R. & Chmielewski, J. 2002, JAChS, 124, 6808

Kashefi, K. & Lovely, D. R. 2003, Sci, 301, 934

Lee, D. H., Granja, J. R., Martinez, J. A., Severin, K. & Ghadiri, M. R. 1996, Natur, 382, 525

Levy, M., Miller, S. L., Brinton, K. & Bada, J. L. 2000, Icar, 145, 609

Miyakawa, S., Yamanashi, H., Kobayashi, K., et al. 2002, PNAS, 99, 14629

Niles, D. H. 2000, Extremophiles, 4, 77

Norlund, K. L. I., Southam, G., Tyliszcak, T., et al. 2009, EnST, 43, 8781

Pick, U. 1999, in Enigmatic Microorganisms and Life in Extreme Environments, ed. J. Seckbach, (Dordrecht: Kluwer), 467

Plaxco, K. W. & Gross, M. 2006, Astrobiology: A Brief Introduction (Baltimore, MD: Johns Hopkins Univ. Press)

Rode, B. M., Fitz, D. & Jakschitz, T. 2007, ChBio, 4, 2674
Rode, B. M. & Schwendinger, M. G. 1990, OLEB, 20, 401
Rothschild, L. J. & Mancinelli, R. L. 2001, Natur, 409, 1092
Schwendinger, M. G. & Rode, B. M. 1989, Anal. Sci., 5, 1377
Tauler, R. & Rode, B. M. 1990, Inorg. Chim. Acta, 173, 93
Trinks, H., Schroder, W. & Biebricher, C. K. 2005, OLEB, 35, 429
Watson, J. D. 2001, The Double Helix: A Personal Account of the Discovery of the Structure of DNA (New York: Touchstone)
Yao, S., Ghosh, I., Zutshi, R. & Chmielewski, J. 1998, Natur, 396, 447

Richard N Boyd and Michael A Famiano

Appendix A

SNAAP Model Mathematics

In writing this book, we have tried to avoid an overwhelming amount of equations. Except for the occasional multiplication or reaction, this book is free from the thorny nest of mathematics.

However, we also recognize that some readers want to know how we arrived at our conclusions along with the details of the mathematics behind our models. For this reason, we supply this appendix with the details of our model. Here, we will describe the definition of the shielding tensor as well as a description of the mathematics used to determine the time-dependent change in the spin states of the nitrogen nucleus in a chiral molecule. If you are content with the conceptual understanding presented in the chapters of this book, you do not need to read this appendix. If, however, you would like to see some of the details that make this model work, then this appendix is for you.

In this appendix, we assume that you are familiar with calculus, differential equations, tensor algebra, and vectors.

A.1 The Shielding Tensor

The shielding tensor is what describes the change in magnetic field at the nucleus due to the orbital electrons in a molecule. All of the physics is contained within this tensor. In a molecule, the shielding tensor defines the reduction in the external magnetic field at a specific nucleus N within that molecule (Buckingham 2004, 2006):

$$\Delta B_{\alpha}^{(N)} = -\sigma_{\alpha\beta}^{(N)} B_{\beta}^{0}. \tag{A.1}$$

The external magnetic field is $\boldsymbol{B_0}$, and the change in field is then $\Delta \boldsymbol{B}^{(N)}$. This reduction is caused by the motion of the electron orbitals. Since the magnetic field is a vector, a tensor is necessary to describe the change in the field at the nucleus. We are adopting the shorthand tensor notation throughout this appendix:

$$\sigma_{\alpha\beta}^{(N)}B_{\beta}^{0} \equiv \left(\sum_{\beta} \sigma_{\alpha\beta}^{(N)}B_{\beta,0}\right)_{\alpha}, \tag{A.2}$$

where the subscript α represents the three components of the magnetic field vector (x, y, z). In this notation, matched subscripts implicitly mean a summation over that subscript.

Because the magnetic field at the nucleus changes, the magnetic shielding tensor also describes the change in the nuclear magnetic moment of a molecule due to the change in electronic orbital motion,

$$\Delta m_{\alpha}^{(N)} = -\sigma_{\alpha\beta}m_{\beta}^{(N)}. \tag{A.3}$$

In the presence of an external electric field, there is a second-order effect on the shielding tensor. Then the tensor, $\sigma_{\alpha\beta}^{(N)'}$, is

$$\begin{aligned} \Delta B_{\alpha}^{(N)} &= -\,\sigma_{\alpha\beta}^{(N)}B_{\beta}^{0} - \sigma_{\alpha\beta\gamma}^{(N)}B_{\beta}^{0}E_{\gamma}^{0} \\ \rightarrow \sigma_{\alpha\beta}^{(N)'} &= \sigma_{\alpha\beta}^{(N)} + \sigma_{\alpha\beta\gamma}^{(N)}E_{\gamma}^{0}, \end{aligned} \tag{A.4}$$

where the prime indicates the tensor corrected for the perturbation from the external electric field. Elements of the third-rank tensor $\sigma_{\alpha\beta\gamma}$, known as the nuclear magnetic shielding polarizability, are defined by

$$\sigma_{\alpha\beta\gamma} = \frac{\partial \sigma_{\alpha\beta}^{(N)}(E)}{\partial E_{\gamma}}. \tag{A.5}$$

We find this by computing the second-rank shielding tensor in the presence of an electric field. For small external fields, if we assume a linear approximation for the derivative, the differences in $\sigma_{\alpha\beta}^{(N)'}$ can result in an estimate for $\sigma_{\alpha\beta\gamma}$,

$$\begin{aligned} \sigma_{\alpha\beta\gamma} &= \lim_{E_{\gamma}\to 0} \frac{\sigma_{\alpha\beta}(E_{\gamma}) - \sigma_{\alpha\beta}(-E_{\gamma})}{E_{\gamma} - (-E_{\gamma})} \\ &= \lim_{E_{\gamma}\to 0} \frac{\sigma_{\alpha\beta}(E_{\gamma}) - \sigma_{\alpha\beta}(-E_{\gamma})}{2E_{\gamma}}. \end{aligned} \tag{A.6}$$

We can also define the isotropic part of the shielding tensor, $\sigma^{(N)}$, which is an average over the axes of the system in an isotropic medium. It provides an average effect of the tensor. These are the diagonal elements of the tensor:

$$\sigma^{(N)} = \frac{1}{3}\sum_{i} \sigma_{ii}. \tag{A.7}$$

We can define a similar value for the polarizability. In this case, the tensor is rank 3 (three-dimensional), so it is like having a tensor for every direction of the electric field. The isotropic part of the polarizability is

$$\overline{\sigma^{(1)}}^{(N)} = \frac{1}{6}\sum_{\alpha\beta\gamma} \sigma^{(N)}_{\alpha\beta\gamma}\varepsilon_{\alpha\beta\gamma}. \tag{A.8}$$

The shift in the electric dipole moment $\Delta\boldsymbol{\mu}_E^{(N)}$, magnetic field $\Delta\boldsymbol{B}^{(N)}$, and magnetic moment $\Delta\boldsymbol{m}^{(N)}$ of a nucleus in the presence of external electric and magnetic fields ($\boldsymbol{E}^{(0)}$ and $\boldsymbol{B}^{(0)}$, respectively; Buckingham 2006) can then be computed in an isotropic medium:

$$\Delta\boldsymbol{\mu}_E^{(N)} = -\overline{\sigma^{(1)}}^{(N)}\boldsymbol{m}^{(N)} \times \boldsymbol{B}^{(0)}, \tag{A.9}$$

$$\Delta\boldsymbol{B}^{(N)} = -\sigma^{(N)}\boldsymbol{B}^{(0)} - \overline{\sigma^{(1)}}^{(N)}\boldsymbol{B}^{(0)} \times \boldsymbol{E}^{(0)}, \tag{A.10}$$

$$\Delta\boldsymbol{m}^{(N)} = -\sigma^{(N)}\boldsymbol{m}^{(N)} + \overline{\sigma^{(1)}}^{(N)}\boldsymbol{m}^{(N)} \times \boldsymbol{E}^{(0)}. \tag{A.11}$$

Here, $\boldsymbol{m}^{(N)}$ is the magnetization for a single nucleus. From the above, it can be seen that there is also an effect on the electric dipole moment as the shielding tensor relates to the motion of the electrons, and a shift in electron distribution also creates a shift in the electric dipole moment.

A.2 Molecules in Magnetic Fields

We can use the derivation of the previous section to discuss effects from the motion of molecules in magnetic fields. This is a well-known effect that comes directly from Maxwell's equations. Electric fields in the molecular reference frame are created via the translational (or motional) Stark effect (Rosenbluh et al. 1977; Panock 1980; Zarnstorff 1997). Electrostatic forces on electrons and nuclei behave in the same way in an external electric field in a nonmoving reference frame. In addition to the molecular energy from the magnetic dipole moment in the presence of a magnetic field, there is also a term from the presence of the molecular dipole moment in the presence of the induced electric field. This energy term is

$$\Delta E = \boldsymbol{\mu}_E \cdot \boldsymbol{v} \times \boldsymbol{B}^{(0)}, \tag{A.12}$$

where $\boldsymbol{\mu}_E$ is the molecular electric dipole moment. A molecule moving in a magnetic field will experience an electric field in its own rest frame equivalent to $\boldsymbol{E}_{TS} \equiv \boldsymbol{v} \times \boldsymbol{B}$. This is a direct result of Maxwell's equations.

For chiral molecules moving in an external magnetic field, the induced electric field in the molecular rest frame thus has two effects. In Equation (A.5), it is seen that the force on the electrons and nuclei in molecules of different shapes results in a reconfiguration of the orbitals' wave functions with respect to the nuclei. This changes the shielding tensor such that the magnetic properties of the molecule change. From Equation (A.9), the reconfiguration of the electric dipole moment in the molecule causes a shift in the energy splittings of the molecule, analogous to the hyperfine splitting in the Zeeman effect.

From Equations (A.10) and (A.11), the shift in magnetic field at the nucleus and its contribution to the total magnetic moment is, in the case of a translation of constant velocity in a uniform magnetic field,

$$\begin{aligned}\Delta \boldsymbol{B}^{(N)} = &- \sigma^{(N)}\boldsymbol{B}^{(0)} \\ &- \overline{\sigma^{(1)}}^{(N)}\boldsymbol{B}^{(0)} \times (\boldsymbol{v} \times \boldsymbol{B}^{(0)}),\end{aligned} \tag{A.13}$$

$$\begin{aligned}\Delta \boldsymbol{m}^{(N)} = &- \sigma^{(N)}\boldsymbol{m}^{(N)} \\ &+ \overline{\sigma^{(1)}}^{(N)}\boldsymbol{m}^{(N)} \times (\boldsymbol{v} \times \boldsymbol{B}^{(0)}).\end{aligned} \tag{A.14}$$

In addition to the effects from the chiral dependence of the shielding tensor, the electric dipole moments, with reversed sign under a chiral transformation, have been shown to exhibit a temperature dependence in the presence of magnetic fields (Buckingham 2014, 2015). This contribution comes from the temperature-dependent orientation of the mean electric polarization in the external electric field, just like the temperature dependence of spin alignment in a magnetic field. The alignment of the chirality-dependent molecular electric dipole moment creates magnetic shielding that depends on the temperature and the electric field. This alignment can be significant as the molecular electric dipole moment can be strong. We can choose a z-axis that is in the direction of the external magnetic field. For an electric dipole moment $\boldsymbol{\mu}_E$, the shift in the magnetization at the nucleus has been shown to be (Buckingham 2014, 2015)

$$\begin{aligned}\Delta_T \boldsymbol{M} = &\frac{1}{6kT}\Big[(\sigma_{xy} - \sigma_{yx})\mu_{E,\,z} \\ &+ (\sigma_{yz} - \sigma_{zy})\mu_{E,\,x} + (\sigma_{zx} - \sigma_{xz})\mu_{E,\,y}\Big]\boldsymbol{M} \times \boldsymbol{E}.\end{aligned} \tag{A.15}$$

In the above equation, M is the net magnetization, i.e., $M = m$ times the number of molecules. In a static electric field, the above is a result of the mean electrical polarization of the molecule oriented in a static external electric field. In this case, the anisotropic components of the shielding tensor times the electric dipole moment vector result in a change in magnetization. This nonisotropic effect can be larger than the isotropic effects of the previous section by several orders of magnitude.

A.3 Magnetization and Spin Alignment in External Fields

Now that we know that nuclear magnetization can vary in external electric and magnetic fields and that this depends on molecular chirality, we can begin to evaluate how certain molecular states may be destroyed over others.

Let us assume that there are two spin states for a nitrogen nucleus in a magnetic field. We can then define the magnitude of the total magnetization as

$$M \equiv (N_+ - N_-)\mu_B, \tag{A.16}$$

where $N_\pm$ is the number of nuclei with spin aligned parallel/antiparallel to the magnetization vector shown in Figure 6.6, and μ_B is the nuclear magnetic moment. Using the value for ^{14}N, the thermal-equilibrium magnetization and populations $N_\pm$ for the D- and L- forms of an amino acid are

$$(N_+ - N_-)_{\mathrm{D,\,L}}\mu_B = \sqrt{M^2 + \Delta M^2_{\mathrm{D/L}}} \equiv M_{eq}. \tag{A.17}$$

Here, the value of ΔM is defined in a way similar to Equation (A.13). Because the sign of $\Delta \boldsymbol{M}$ is opposite for each chirality, the equilibrium magnetization $\boldsymbol{M}_{eq}$ has equal magnitudes for each chirality.

In an antineutrino flux, specific spin states are then selectively destroyed by the nuclear interaction with the antineutrinos. We can then estimate the time evolution of the component of magnetization parallel to the external field. The magnetization is changed by the selective destruction of spin-aligned nuclei relative to anti-spin-aligned nuclei in the magnetic field. It is also changed by the thermalization of the spins (the relaxation). For one particular chirality, the magnitude of M evolves as

$$\frac{dM}{dt} = \left(\frac{dM}{dt}\right)_\nu + \left(\frac{dM}{dt}\right)_T. \tag{A.18}$$

The first term above is a result of the destruction of specific spin states of the nitrogen nucleus by the antineutrinos. The second term is the result of the redistribution of the spin states due to thermal relaxation. The first term destroys N nuclei in a particular spin state, while the second term reorganizes spin states. For the second term above, the total number of spin states, $N_+ + N_-$, is conserved. As a result, $dN_+/dt = -dN_-/dt$. The spin states can be written in terms of the thermal relaxation and total magnetization:

$$\mu_B\frac{dN_+}{dt} = -\mu_B\frac{dN_-}{dt} = \frac{1}{2}\left(\frac{dM}{dt}\right)_T, \tag{A.19}$$

where the "D" and "L" subscripts have been dropped for readability and are understood.

The spin states for a particular chirality can then be written for the L-enantiomer as

$$\begin{aligned}\frac{dN_+}{dt} &= -\,R_pN_+ + \frac{1}{2\mu_B}\left(\frac{dM}{dt}\right)_T\\ \frac{dN_-}{dt} &= -\,R_aN_- - \frac{1}{2\mu_B}\left(\frac{dM}{dt}\right)_T,\end{aligned} \tag{A.20}$$

where $R_{p(a)}$ is the antineutrino-capture rate for antineutrino spin that is (anti)-parallel to the nuclear spin. (Note the change in sign in the second term as the thermalization results in the destruction of one spin state while creating another.) From Figure 6.6, it is seen that Equation (A.20) applies to the L-enantiomer. For rates of the D-enantiomer, R_p and R_a are exchanged. In this formulation, it can be shown that $R_a > R_p$.

Utilizing the thermal relaxation relationship for a nuclear spin in a magnetic field, Equation (A.20) becomes

$$\begin{aligned} \frac{dN_+}{dt} &= -R_p N_+ + \frac{1}{2\mu_B}\left(\frac{M_{eq} - M(t)}{T_1}\right) \\ \frac{dN_-}{dt} &= -R_a N_- - \frac{1}{2\mu_B}\left(\frac{M_{eq} - M(t)}{T_1}\right), \end{aligned} \tag{A.21}$$

where T_1 is the longitudinal relaxation time for the molecular configuration in the medium. Substituting the magnetization into the above equation results in a set of coupled differential equations:

$$\begin{aligned} \frac{dN_+}{dt} &= -R_p N_+ + \frac{M_{eq}}{2\mu_B T_1} - \frac{1}{2T_1}N_+ + \frac{1}{2T_1}N_- \\ &= -R_p N_+ + \frac{f_+ N - f_- N}{2T_1} - \frac{1}{2T_1}N_+ + \frac{1}{2T_1}N_- \\ &= -R_p N_+ + \frac{\Delta f N}{2T_1} - \frac{1}{2T_1}N_+ + \frac{1}{2T_1}N_- \\ \frac{dN_-}{dt} &= -R_a N_- - \frac{M_{eq}}{2\mu_B T_1} + \frac{1}{2T_1}N_+ - \frac{1}{2T_1}N_- \\ &= -R_a N_- - \frac{f_+ N - f_- N}{2T_1} + \frac{1}{2T_1}N_+ - \frac{1}{2T_1}N_- \\ &= -R_a N_- - \frac{\Delta f N}{2T_1} + \frac{1}{2T_1}N_+ - \frac{1}{2T_1}N_-, \end{aligned} \tag{A.22}$$

where $f_{+(-)}$ is the fraction of nuclei in the spin up (down) state in thermal equilibrium with the external magnetic field and $\Delta f \equiv f_+ - f_-$. The total nuclear population is $N = N_+ + N_-$.

In thermal equilibrium, we can define $f_{+/-}$ as

$$\begin{aligned} f_- &= \frac{\exp[-\Delta E/kT]}{1 + \exp[-\Delta E/kT]} \\ f_+ &= \frac{1}{1 + \exp[-\Delta E/kT]}, \end{aligned} \tag{A.23}$$

where ΔE is the difference in energy for a nucleus that is aligned with the magnetic field and that aligned against the field:

$$\Delta E = 2\mu_B B. \tag{A.24}$$

The above equation can be written as a homogeneous first-order differential equation in vector form:

$$\dot{\boldsymbol{N}} = \lambda \begin{pmatrix} -\alpha & \delta \\ \epsilon & -\beta \end{pmatrix} \boldsymbol{N}, \tag{A.25}$$

where the following are defined:

$$\begin{aligned}
\alpha &\equiv \begin{cases} 2T_1R_p + 1 - \Delta f & \text{L-states} \\ 2T_1R_a + 1 - \Delta f & \text{D-states} \end{cases} \\
\beta &\equiv \begin{cases} 2T_1R_a + 1 + \Delta f & \text{L-states} \\ 2T_1R_p + 1 + \Delta f & \text{D-states} \end{cases} \\
\delta &= 1 + \Delta f \\
\epsilon &= 1 - \Delta f \\
\lambda &\equiv \frac{1}{2T_1},
\end{aligned} \tag{A.26}$$

and the vector of spin-state numbers (N_+, N_-) in Equation (A.25) is

$$\boldsymbol{N} = \begin{pmatrix} N_+ \\ N_- \end{pmatrix}. \tag{A.27}$$

The above set of equations has the solutions:

$$\boldsymbol{N}_\chi = \begin{pmatrix} c_1 & \dfrac{\lambda}{r_- + \lambda\alpha} c_2 \\ \dfrac{r_+ + \lambda\alpha}{\lambda} c_1 & c_2 \end{pmatrix} \begin{pmatrix} e^{r_+ t} \\ e^{r_- t} \end{pmatrix}. \tag{A.28}$$

In the above, the chirality-dependent terms are $r_{+/-}$ and α. The values of c_1 and c_2 are determined by the boundary conditions, and the following are defined:

$$\begin{aligned}
r_\pm &\equiv \frac{\lambda}{2}\{-(\alpha + \beta) \pm [(\alpha - \beta)^2 + 4\delta\epsilon]^{1/2}\} \\
&= -\lambda\Big[T_1(R_p + R_a) + 1 \\
&\quad \mp \sqrt{(T_1(R_p - R_a))^2 + 1 - 2\Delta f T_1(R_p - R_a)}\Big].
\end{aligned} \tag{A.29}$$

In the above, $r_\pm < 0$, which means that individual spin-state numbers are decaying in time owing to destruction via antineutrino interactions. The above relationship is for L-enantiomers; for D-enantiomers, the subscripts are reversed.

The total abundance of a chiral state is obtained by summing the components of the spin-state vector from Equation (A.28):

$$N_\chi = N_{+,\chi} + N_{-,\chi}. \tag{A.30}$$

Given the above, the enantiomeric excess of a particular state can now be computed in the SNAAP model,

$$ee = \frac{(N_{+,\,\mathrm{L}} + N_{-,\,\mathrm{L}}) - (N_{+,\,\mathrm{D}} + N_{-,\,\mathrm{D}})}{N_{+,\,\mathrm{L}} + N_{-,\,\mathrm{L}} + N_{+,\,\mathrm{D}} + N_{-,\,\mathrm{D}}}. \tag{A.31}$$

As an example, we evaluate *ee* by assuming the boundary conditions $N_+(t=0) = N_-(t=0) = N/2$, where N is the total population of molecular states. Using the same boundary conditions for both chiral states gives $N_\mathrm{D}(t=0) = N_\mathrm{L}(t=0) = N/2$. One can then set individual combinations of spin and chirality $N_{+,\,\mathrm{D}} = N_{-,\,\mathrm{D}} = N_{+,\,\mathrm{L}} = N_{-,\,\mathrm{L}} = N/4$. These initial conditions result in $M(t=0) = ee(t=0) = 0$. With these conditions, the constants c_1 and c_2 are evaluated for each set of chiral equations:

$$\begin{aligned} c_{1,\,\chi} &= \frac{N}{4}\left(\frac{r_{-,\,\chi} + \lambda\alpha_\chi - \lambda\delta}{r_{-,\,\chi} - r_{+,\,\chi}}\right) \\ c_{2,\,\chi} &= \frac{N}{4}\left(\frac{r_{+,\,\chi} + \lambda\alpha_\chi - \lambda\delta}{\lambda\delta}\right)\left(\frac{r_{-,\,\chi} + \lambda\alpha_\chi}{r_{+,\,\chi} - r_{-,\,\chi}}\right), \end{aligned} \tag{A.32}$$

where the subscripts have been explicitly inserted to indicate values that are chirality dependent.

Equation (A.28) shows that one spin–chiral combination is preferentially destroyed. Figure 6.6 shows that the antineutrino spin alignment with the parallel (N_+) nitrogen spin alignment in the L-enantiomer is less than 90° while it is greater than 90° for the D-enantiomer. This means that the D-enantiomer is preferentially destroyed in this orientation for meteoroids that are moving in a net positive radial direction. This is because $\Delta \boldsymbol{M}_\mathrm{L}$ points in the direction of the velocity vector in this model. For meteoroids following an elliptical or hyperbolic trajectory, it is quite possible for a net *ee* to result as the meteoroid passes the star. If its path is at any non-equatorial inclination, the magnetic field strength in the meteoroid's reference frame would be different for the approach and departure from the star. The different field strengths would result in a different destruction rate of L- and D-enantiomers, producing a net *ee*.

The antineutrino interaction rate on ^{14}N is a parameter of this model. Here, the rate of antineutrino interactions for antineutrinos spin-aligned parallel to the nitrogen spin is taken to be a factor of the inverse of the relaxation time $R_p = f\lambda = f/(2T_1)$. Figure 6.6 shows that the nuclear spins are, on average, nearly perpendicular to the antineutrino velocity vectors. The ratio of rates for spins parallel and antiparallel to the ^{14}N spin is

$$\frac{R_p}{R_a} = \frac{1 - \cos\Theta_p}{1 - \cos\Theta_a} = \frac{1 - \sin\phi}{1 + \sin\phi} \approx \frac{1 - \phi}{1 + \phi}. \tag{A.33}$$

A larger $\Delta M/M$ can result in a larger ratio of rates, resulting in a larger *ee*.

References

Buckingham, A. 2004, CPL, 398, 1
Buckingham, A. & Fischer, P. 2006, CP, 24, 111
Buckingham, A. D. 2014, JChPhy, 140, 011103
Buckingham, A. D., Lazzeretti, P. & Pelloni, S. 2015, MolPh, 113, 1780
Panock, R., Rosenbluh, M., Lax, B. & Miller, T. A. 1980, PhRvA, 22, 1041
Rosenbluh, M., Miller, T. A., Larsen, D. M. & Lax, B. 1977, PhRvL, 39, 874
Zarnstorff, M. C., Levinton, F. M., Batha, S. H. & Synakowski, E. J. 1997, PhPl, 4, 1097

Richard N Boyd and Michael A Famiano

Appendix B

True and False Chirality

We discussed previously how external influences can select the chirality of a system (either for creation, conversion, or destruction). A number of theories are possible in which some sort of asymmetric external influence can change the net resultant chirality of the system.

B.1 Parity Transformations

Vectors are defined as rank-1 tensors, and the components of this tensor define the magnitude and direction of the vector. We have mentioned a few specific vectors such as velocity and angular momentum (or spin). By convention, the spin vector direction is perpendicular to the rotational plane of the spinning object.

However, vectors can also be defined by how they behave under a parity transformation and a time-reversal transformation. Perhaps the simplest possible definition of parity is whether something is positive or negative. We also use the terms even and odd to define the parity of a vector or system of vectors. A parity transformation simply replaces the coordinate system with one in which the components point in the negative direction. That is, $(x, y, z) \rightarrow (-x, -y, -z)$. A parity transformation can also be accomplished by taking the mirror image of a coordinate system, $x \rightarrow -x$, followed by a rotation of the coordinate system about the x-axis. Many times, one sees parity transformations expressed as a reflection across one coordinate plane followed by a rotation of 180° about the reflected axis. One can rotate the coordinate system or the object. In any case, the coordinates are swapped.

Consider the situation shown in Figure B.1. Here, we show an arrow and a coordinate system. For simplicity, let us assume that the arrow lies entirely in the yz plane as indicated in the figure. Under a parity transformation (conventionally indicated by "P"), we reflect the coordinate system over the xz plane. Next, we can

doi:10.1088/978-0-7503-1993-5ch12

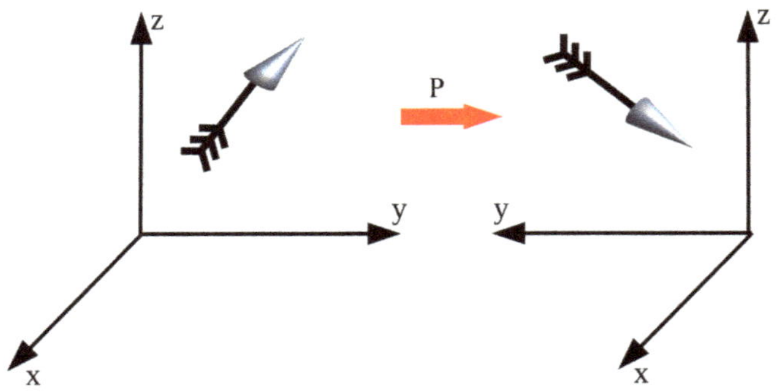

Figure B.1. Parity transformation of an arrow. To accomplish this, we reflected the coordinate system about the *xz* plane and rotated the arrow about the *y*-axis by 180°. (Note that we have also translated the arrow, which is inconsequential for this example as vectors can be translated as we wish.)

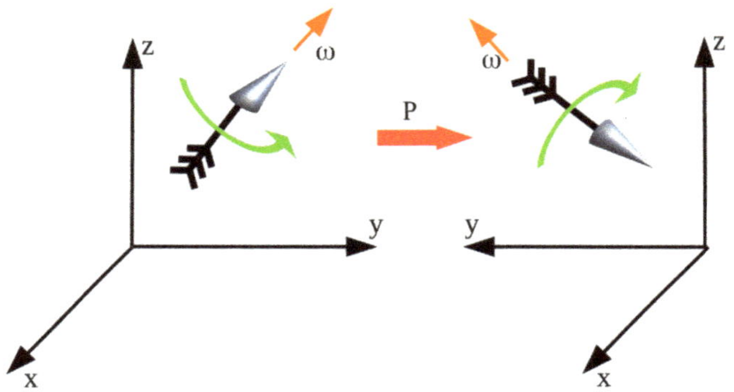

Figure B.2. Parity transformation of an arrow that is spinning as it flies. We accomplish the parity transformation in the same way as we do for Figure B.1. The direction of rotation is shown in the figure.

either rotate the coordinate system around the *y*-axis, or we can accomplish the same thing by rotating the arrow around the *y*-axis. We choose to rotate the arrow in this case. By doing all of these, the direction of the arrow changes. Instead of pointing in the positive *y*- and *z*-directions, the arrow now points in the negative *y*- and *z*-directions. Other vectors behave this way, including velocity, momentum, radius, and force.

Now, let us assume that along with flying through the air, the arrow is spinning about its long axis as shown in Figure B.2. Note that because of the arrow's spin, we now have two vectors involved: the direction to where arrow is pointing and the angular momentum vector, and we can treat them independently. If we perform a reflection followed by a rotation as above, the arrow's configuration is the one shown on the right-hand side of the figure. (If you have a difficult time visualizing this, try it with a pencil and draw a mark on the pencil to indicate rotation direction, making sure to change the direction of the mark if you reflect the pencil.)

As before, we see that the direction of the arrow changes sign. However, the angular momentum vector of the arrow does not. We can verify this with the right-hand rule.

It would appear that the arrow's direction vector and angular momentum vector do not behave in the same way under a parity transformation. The angular momentum vector is invariant—or even—under a parity transformation, while the direction vector is not; it is *odd* under a parity transformation. The angular momentum vector is referred to as an axial vector or a pseudovector. Other axial vectors include angular velocity, torque, and magnetic field. The direction vector is referred to as a polar vector. Polar vectors change sign under a parity transformation, while axial vectors do not.

We can also imagine scalar and vector quantities that result from vector operations. In the case of dot products, two vectors of the same type result in a value that does not change sign under a parity transformation. The same is true for cross products. An example of a scalar formed by two polar vectors is the work done by a force, $\boldsymbol{F} \cdot \boldsymbol{r}$. Under a parity transformation, the force and distance both change sign, which means the resultant dot product is still positive.

Vector quantities of two different types, on the other hand, have resultant vector operations that do change sign under a parity transformation because only one vector changes sign in the operation. An example of this used in this book is a particle's helicity, which is defined as

$$h = \frac{\boldsymbol{\sigma} \cdot \boldsymbol{p}}{|\boldsymbol{\sigma} \cdot \boldsymbol{p}|}, \tag{B.1}$$

where $\boldsymbol{\sigma}$ is the spin of the particle, and $\boldsymbol{p}$ is the particle's momentum. The spin is an axial vector that does not change sign under a parity transform, but the momentum does. Thus, a parity transformation reverses the sign of the helicity. One can imagine the arrow in Figure B.2 as a momentum vector and the spin changing direction with respect to the momentum. A scalar value that changes sign under a parity transformation is referred to as a pseudoscalar.

B.2 Time-reversal Transformations

Another type of transformation is the time-reversal transformation. A time-reversal transformation is implemented by reversing the sign of the time quantities, $t \to -t$, $\Delta t \to -\Delta t$, and $\partial t \to -\partial t$. Some have said that a time-reversal transformation is like running the movie backwards when observing physical phenomena. Consider the arrow of Figure B.2. If the arrow is just sitting still, its position does not change in time. Since the position is independent of time, position is even under time reversal. It does not change sign.

However, if the arrow is moving forward with some velocity $\boldsymbol{v}$ and some angular velocity $\boldsymbol{\omega}$ as shown in Figure B.3, the orientation or position at any given time would not change under time reversal, but both the velocity and the angular velocity would change direction. Under this transformation, the arrow would move in the opposite direction, and it would spin in the opposite direction.

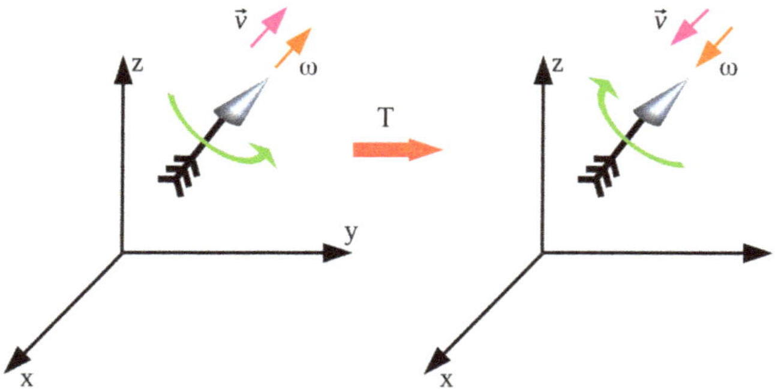

Figure B.3. Time-reversal transformation for an arrow that is moving and spinning at the same time.

The time-reversal transformation comes about because of the change in sign of the time value:

$$\begin{aligned} \boldsymbol{v} &= \frac{d\boldsymbol{r}}{dt} \xrightarrow{T} \frac{d\boldsymbol{r}}{-dt} = -\boldsymbol{v} \\ \boldsymbol{\omega} &= \frac{d\boldsymbol{\theta}}{dt} \xrightarrow{T} \frac{d\boldsymbol{\theta}}{-dt} = -\boldsymbol{\omega}. \end{aligned} \tag{B.2}$$

Vectors that depend on time to odd order change sign under time reversal, while vectors that depend on time to even order do not change sign under time reversal.

B.3 Magnetic Fields

Magnetic fields are worth mentioning here as they comprise a significant portion of the descriptions in this book. Magnetic fields are interesting under parity and time-reversal transformations. They can be potentially confusing, and much of this confusion can be eliminated by understanding how magnetic fields behave and where they come from.

In terms of a parity transformation, it is useful to think of the magnetic field in terms of the vector potential, $\boldsymbol{A}$. The vector potential is a vector field, as it has a vector quantity assigned to every point in space. The vector potential can be related to the magnetic field via the curl of the potential:

$$\boldsymbol{B} = \nabla \times \boldsymbol{A}. \tag{B.3}$$

Under a parity transformation, the vector potential changes sign. However, the curl also changes sign:

$$\nabla = \left(\frac{\partial}{\partial x}, \frac{\partial}{\partial y}, \frac{\partial}{\partial z}\right) \xrightarrow{P} \left(\frac{\partial}{-\partial x}, \frac{\partial}{-\partial y}, \frac{\partial}{-\partial z}\right) = -\nabla. \tag{B.4}$$

Thus, the magnetic field is even under a parity transformation.

One might think that the magnetic field is even under time reversal as well. However, it is useful to consider the actual source of magnetic fields from Maxwell's equations. Magnetic fields come from time-varying electric fields. That includes

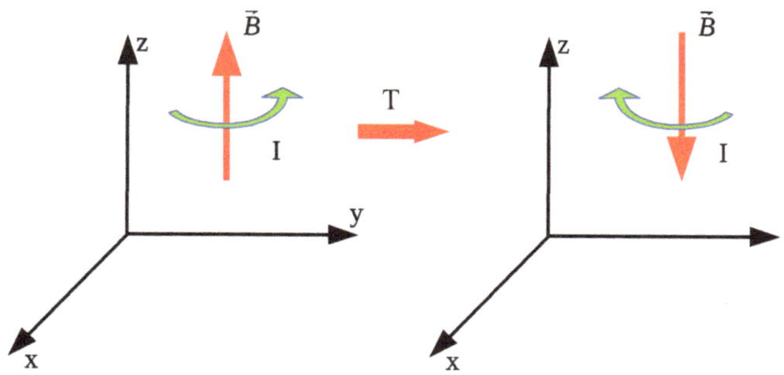

Figure B.4. A magnetic field created by a current loop. Under a time-reversal transformation, the current loop changes direction, thus also changing the direction of the magnetic field.

moving charges as well. Under time reversal, moving charges change direction, as do time-varying electric fields. Thus, magnetic fields are odd under time reversal. Maxwell's equations provide an explanation of the source of magnetic fields:

$$\nabla \times \boldsymbol{E} = -\frac{d\boldsymbol{B}}{dt}. \tag{B.5}$$

The left-hand side of the equation is time-even. However, the right-hand side is time-odd because of the time differential in the denominator. Thus, a time-reversal transformation will reverse the sign of magnetic fields.

Consider Figure B.4. In this figure, a magnetic field produced by a current loop is shown. Under a time-reversal transformation, the direction of the current changes direction. As a result, the direction of the magnetic field changes. This concept also applies to how magnetic fields are created in matter. Microscopically, magnetic fields are created in magnetic domains, which result from electron currents in atoms. Reversing the time reverses the electron current, and permanent magnetic fields are also reversed.

B.4 Chiral Transformations

Is a transformation or process truly chiral? That is, does a process truly select or target one chiral state over another? Another way of thinking about this question is that a transformation that selects one chiral state must select the opposite chiral state if coordinates are reversed. However, truly chiral processes must not be changed if time is reversed. That is, truly chiral processes are space dependent, but not time dependent (Barron 2008, 2013). This is because true chirality is also a space-dependent process, but not a time-dependent one.

Consider the simple example of neutrinos and antineutrinos as shown in Figure B.5. For now, just consider antineutrinos, which have a positive helicity; they are right-handed. What will happen if we consider a parity transformation or a time-reversal transformation as shown in Figure B.5? The neutrino has both a velocity and an angular momentum. Under a parity transformation, the velocity changes sign, but the angular momentum does not. Thus, a parity transformation switches the antineutrino's helicity. Under a time-reversal transformation, both the velocity

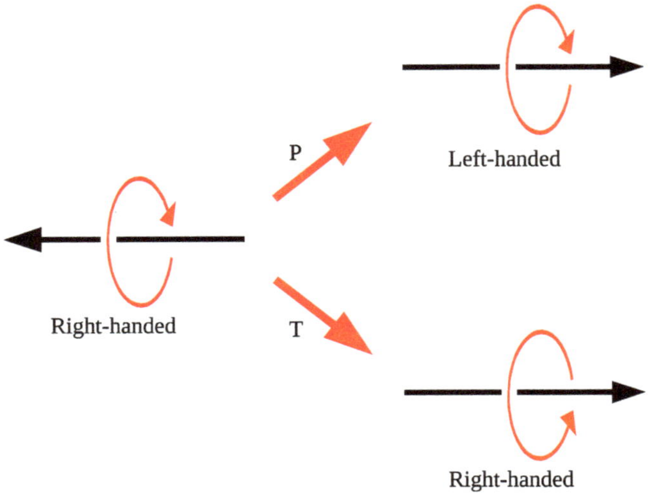

Figure B.5. An antineutrino after undergoing a parity transformation, P, and a time-reversal transformation, T.

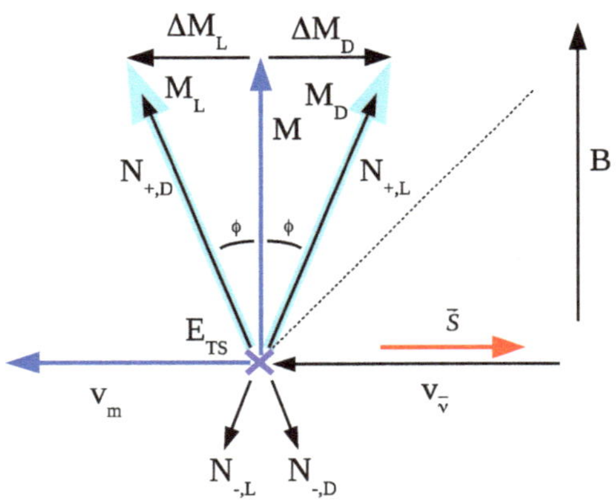

Figure B.6. The magnetochiral effects in the SNAAP model under a parity transformation.

and the angular momentum change sign. Thus, the antineutrino is still right-handed under a time-reversal transformation. The antineutrino satisfies both conditions for a truly chiral phenomenon. It is indeed a truly chiral particle.

Is the SNAAP model truly chiral? To evaluate this, we examine the vectors that describe the magnetochiral effects of the SNAAP model in Figure 6.6. The parity transformation of this figure is shown in Figure B.6. (Here, we show the figure with the antineutrino velocity vector at an angle $\theta = 0$.) Under a parity transformation, neither the magnetic field nor the antineutrino spin changes direction, as shown. As a result, the unperturbed magnetization $\boldsymbol{M}$ does not change sign. However, all other vectors change sign, including the induced electric field, which is now into the page. Because of this, the shifts in magnetization $\Delta\boldsymbol{M}_{\mathrm{L}}$ and $\Delta\boldsymbol{M}_{\mathrm{D}}$ also change sign. The final result is that the effect of this model is reversed, with the antineutrino spin

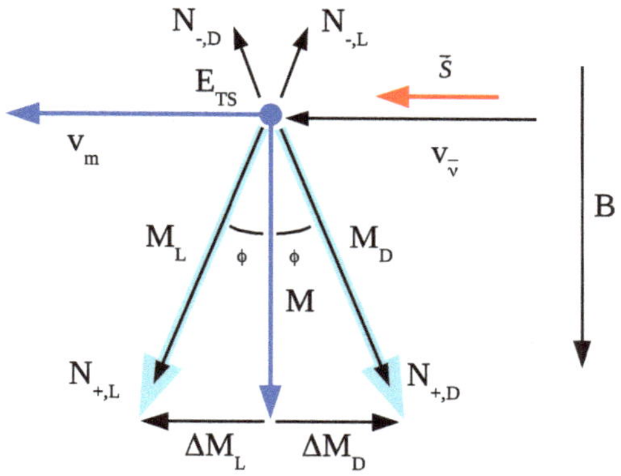

Figure B.7. The magnetochiral effects in the SNAAP model under a time-reversal transformation.

parallel to $\Delta \boldsymbol{M}_{\rm D}$ instead of antiparallel, and left-handed amino acids are preferentially destroyed instead of right-handed amino acids as in Figure 6.6.

Under a time-reversal transformation of Figure 6.6, the velocities, spins, and magnetic field change sign. As a result, the unperturbed magnetization changes sign. However, the induced electric field does not change sign. The result of this transformation is shown in Figure B.7.

The shift in magnetization, which is proportional to $\boldsymbol{M} \times \boldsymbol{E}$, changes sign as well. Thus, the antineutrino spin is antiparallel to the right-handed magnetization and parallel to the left-handed magnetization. As a result, right-handed amino acids are still preferentially destroyed under a time-reversal transformation.

Thus, this model is parity-odd and time-even. The effect proposed by this model is a true chiral effect.

It is worth pointing out that chirality selection does not necessarily have to be a true chiral effect (Barron 2013). However, for equilibrium conditions, a true chiral effect is necessary. For a system driven out of equilibrium, true chiral effects may not be necessary. A falsely chiral effect may work, but the requirements may be more difficult to satisfy.

References

Barron, L. D. 2008, SSRv, 135, 187

Barron, L. D. 2013, Rend. Fis. Acc. Lincei, 24, 179

Richard N Boyd and Michael A Famiano

Glossary

Absorption spectra	Spectra resulting from the absorption in a star's surface of the radiation produced inside the star that backlights the surface region.
Adaptive optics	Ability of a telescope to shape its primary mirror over a short timescale in order to minimize the distortion caused by Earth's atmosphere and optimize the resolution.
Adenine	One of the nucleotide bases that constitute DNA.
Algae	Unicellular or multicellular chlorophyll-containing plants occurring in fresh or salt water.
Alpha-amino acids	A class of amino acids that is distinguishable from other amino acids by the location of its different groups. α-amino acids are the ones that are important to Earthly life.
Amino acid	Molecules that are all characterized by having an amino group, NH_2, and a carboxyl group, COOH (see Figure 5.1 for a picture), along with the other atoms that distinguish them from each other. Amino acids are the components of the proteins we rely on for life.
Amino group	NH_2. One of the basic components of all amino acids along with the carboxyl group.
Amplification	Process through which low levels of enantiomerism can be increased to much higher levels, or even to homochirality. This could occur either on a planet or in the cosmos, although the mechanisms in the two environments may be quite different.
Angular momentum	A conserved quantity in physics that is associated with rotational motion.
Anion	Negatively charged ion created by the addition of an extra electron or the removal of a hydrogen nucleus.
Archaea	Single-celled organisms that have no nucleus and constitute one of the three basic groups of archaea, bacteria, or eukaryotes.
Asteroid Belt	A huge system of interplanetary debris that exists between the orbits of Mars and Jupiter.
Astronomical unit	Abbreviated au; this is the mean distance between Earth and the Sun, which is 1.5×10^{13} cm.

doi:10.1088/978-0-7503-1993-5ch13

Asymptotic giant branch stars	Stars that have completed burning the hydrogen in their cores and have evolved from exhibiting the characteristics of that burning mode into exhibiting those of helium burning.
Autocatalysis	Chemical replication schemes by which molecular properties can be enhanced.
Axial vector	A vector that is invariant under a parity transformation. Also referred to as a pseudovector.
Bacteria	Single-celled organisms that, like archaea, have no nucleus and constitute one of the three basic groups—archaea, bacteria, or eukaryote. However, they also differ from archaea in several ways.
Baryons	Particles that are made up of three quarks. Protons and neutrons are baryons; photons, electrons, and neutrinos are not. Pions are also not baryons.
Baryon conservation	One of the laws of physics that says that the number of baryons in any reaction or decay must be conserved.
Beta-decay	Process by which an unstable nucleus emits either an electron and an electron antineutrino or a positron and an electron neutrino, or captures an electron and emits an electron neutrino. In doing so, it will form a more stable nucleus.
Big Bang	The creation event of our universe.
Binding energy	Energy binding a nucleus or molecule, that is, the energy that would be required to break it apart into at least some of its constituents.
Black hole	Final state of a very massive star after it has completed all stages of its stellar evolution. No matter can escape from a black hole. Massive black holes also exist at the centers of galaxies.
Black smokers	Hydrothermal vents in the ocean floor from which a variety of substances pour out.
Bremsstrahlung	Literally, "braking radiation"; produced when a charged particle is accelerated or decelerated
Buckingham effect	Result of the interaction between a nucleus that has a nonzero spin and an external magnetic field. This was originally devised to describe how nuclear magnetic resonance could be used to differentiate between molecules of opposite chirality.
Cambrian explosion	Event in Earth's history 500–600 million years ago when there was an incredible expansion in the diversity of life forms.
Carbonaceous chondrites	Type of meteorite that appears particularly capable of withstanding the high temperatures associated with passage through Earth's atmosphere. Some of these have been found to contain amino acids.
Carboxyl group	COOH; one of the basic components of all amino acids along with the amino group.
Carbon burning	One of the stages of stellar evolution, following helium burning and preceding neon burning.
Catalysis	In catalysis, the "catalyst" nucleus or molecule enables a process in which new nuclei or molecules are formed, and the original nucleus or molecule is returned.
Cation	Positively charged ion created by removing one electron.

Centigrade	One of several temperature scales. On this scale, water freezes at 0 °C and boils at 100 °C.
Cepheid variables	Stars that have an oscillatory output, the frequency of which is related to the absolute luminosity of the star. Cepheids, therefore, have been used as distance indicators.
Chandra X-Ray Observatory	One of the spaceborne observatories designed to detect X-rays. This observatory was named in honor of W. Chandrasekhar.
Charged-current weak interactions	Weak interactions that change from one nucleus to another of nearly equal mass, but with different numbers of protons and neutrons.
Chart of nuclides	The chart of all known nuclides, stable and unstable. This includes all known isotopes of every element.
Chemical evolution (of the Galaxy)	Buildup in time of the elements in the interstellar medium.
Chemical evolution (of biomolecules)	Evolution of the biomolecules from simple amino acids into peptides, and ultimately to the more complex DNA and RNA.
Chirality	Handedness. Chiral molecules have an "opposite," which has the opposite chirality. Chirally opposite molecules cannot be translated in such a way that they are identical, but they do appear identical if one of them is reflected in a mirror.
Chlorophyll	Green coloring of matter that enables the production of carbohydrates by photosynthesis.
Churyumov–Gerasimenko	Comet that was the destination of the European Space Agency mission *Rosetta*.
Circularly polarized light	Light for which the electric field vector appears to rotate (see Figure 5.3). In right circularly polarized light, the electric field vector appears to rotate in a clockwise direction when the light is viewed in the direction of travel, and vice versa.
CNO cycle	Describes the details of the dominant form of hydrogen burning, in which four hydrogen nuclei get fused into a helium nucleus in stars more massive than several solar masses.
Cold dark matter	One of the hypothetical forms of dark matter, in which the particles are assumed to exist at low temperature.
Comet	A celestial object, often quite large (compared to a meteorite, but not to a planet), that is usually composed of ice and dust.
Conservation laws	The laws of physics that govern processes and forms. Examples include conservation of energy and conservation of angular momentum.
Core-collapse supernovae	Supernovae that are produced by massive stars when they have completed all stages of their stellar evolution.
Cosmology	Study of the origin of our universe. This also includes identification of the observable signatures of our universe's birth event.
Coulomb barrier	Barrier resulting from the interaction of two charged particles of the same sign. In particles at low energy, the Coulomb barrier works to prevent the particles from undergoing any interaction other than the electromagnetic interaction, although they can interact via the strong interaction and undergo fusion via quantum mechanical tunneling.

Cross section	Reaction probability for one particle to interact with another. See reaction probability.
Cryophiles	Cold-loving creatures; a class of extremophiles.
Cyanobacteria	Blue-green bacteria or archaea; its cells have reached a relatively high level of sophistication, at least for prokaryotes.
Cyanidium caldarium	A living organism that can tolerate extremely low pH values, that is, very acidic environments.
Cytoplasm	Liquid between the cell walls and the nucleus in eukaryota, and within the cells of non-eukaryota cells.
Cytosine	One of the nucleotide bases constituting DNA.
Dark energy	Entity that appears to cause the expansion of the universe to accelerate.
Dark matter	Form of matter, as yet unknown, that exerts a gravitational force on other types of matter, and so influences the motion of galaxies.
Degeneracy pressure	Pressure created by the Pauli Principle, which states that only one identical particle of a particular type (for example, electrons or neutrons) can occupy a single quantum state. The degeneracy pressure due to electrons is the pressure that maintains the sizes of atoms and of white dwarfs, and that due to neutrons and protons maintains the sizes of atomic nuclei and of neutron stars.
Density functional theory	Theory that allows the electronic structure of molecules to be calculated by fitting wavefunctions to a set of existing functions.
Deuteron	Nucleus of heavy hydrogen, that is, a nucleus that contains one proton and one neutron.
Deoxyribonucleic acid, DNA	One of the chemicals of life. It contains the instructions for assembling proteins within cells and also contains all of the genetic information essential for replication.
Diamagnetic	Having a negative, nonzero magnetic susceptibility. Objects that are diamagnetic will have magnetizations that point in the opposite direction to an external magnetic field. Thus, these objects are repelled by external magnetic fields.
Doppler shift	Shift in the spectral lines of atoms or ions with motion. When atoms or ions are moving away from the observer, the characteristic lines will be "redshifted."
Deinococcus radiodurans	Living organism that can tolerate, via repair mechanisms, very high levels of radiation.
Drake equation	Equation of sorts that categorizes the different factors that would be required for some extraterrestrial civilization to send radio signals that we could interpret as indicating the existence of an advanced civilization.
Dunaliella acidophia	A living organism that can tolerate extremely low pH, that is, highly acidic, environments.
Dust grains	The tiny (of order 10 μm) grains of molecules that form in the interstellar medium.
Efflux pump	Mechanism by which cells rid themselves of toxic entities, for example, zinc, copper, or cobalt.
Electric dipole moment	The electric dipole moment of a molecule is generated by charge distributions in which the positive and negative charges are not distributed evenly in the same space.

Electric field vector	Vector that specifies both the strength and direction of the electric field.
Electric field	Distribution of that physical entity that affects charged particles and, together with the magnetic field, represents photons as they move through space.
Electromagnetic radiation	Radiation that is the result of electric and magnetic fields. The particles of electromagnetic radiation are called photons.
Electromagnetic force/interaction	One of the four basic interactions of physics.
Electron	A light fundamental particle. It has a spin of $\hbar/2$ and a negative charge.
Element	Atoms of a particular element are characterized by the number of protons in their nucleus. For a neutral atom (that is, it is not ionized), the number of electrons the atom has will be equal to its number of protons. Hydrogen, helium, oxygen, carbon, tin, etc. are elements.
Elliptically polarized light	Light that has its electric field vectors somewhat out of phase, but not necessarily 90° out of phase, which would produce circularly polarized light, or with its electric field not necessarily having two equal-strength components.
Enantiomer	One particular "mirror" image of a molecule that is chiral. For example, the L-enantiomer refers to left-handed molecules, while the D-enantiomer refers to right-handed molecules
Enantiomeric	Medium in which the two possible chiral states are not equally populated.
Enantiomeric excess	Amount by which one chirality exceeds the other, divided by their sum. It is usually also expressed as a percentage, so the difference divided by the sum is multiplied by 100.
Endothermic	Refers to reactions that absorb energy, that is, they require energy to make them occur.
Entropy	Measure of the number of available states in a system. Systems with larger entropy have a larger number of possible configurations. Systems with zero entropy have only one single possible configuration.
Enzyme	Proteins that act as catalysts to, among other functions, increase the rates of metabolic reactions.
Erg	Unit of energy. A 93 mile-per-hour fastball has 12 billion ergs of energy.
Eukaryotes	Organisms that are the most complex of the three groups: archaea, bacteria, and eukaryote. Specifically, eukaryotes have a nuclear envelope in which their genetic material is contained, whereas archaea and bacteria have no nucleus.
Europa	One of the moons of Jupiter.
Event horizon	Radius at which escape from a black hole becomes impossible.
Evolution	Natural progression from one entity to another. Specifically, this describes how species change with time.
Excited (nuclear or atomic) state	Allowed state where the nucleus or atom can exist at a higher (less tightly bound) energy than the ground state.
Exothermic	Refers to reactions that give off energy when they occur.

Extraterrestrial	Not of Earth.
Extremophiles	Creatures that prefer living conditions very different from those human beings prefer. These might include temperature extremes, pressure extremes, unusually acidic or alkaline conditions, etc. They might also be able to withstand extremes that humans could not, for example, high radiation levels.
Fahrenheit	One of several temperature scales. On this scale, absolute zero is −459.67 °F, and water freezes at 32 °F and boils at 212 °F.
Faraday's Law	One of Maxwell's four equations for electromagnetism, which states that a changing magnetic flux will produce an electric field.
Fermi paradox	Statement, by Enrico Fermi, about the three possibilities why we have not observed extraterrestrial life.
Fermion	Particle with a quantum mechanical spin of $n/2$, where n is an odd integer. No two fermions can occupy the same quantum state.
Ferromagnetic	Having a large, positive, nonzero magnetic susceptibility. Objects that are ferromagnetic will have a large net magnetization in the presence of an external magnetic field. These objects are strongly attracted to other magnetic objects.
Field	Mathematical construct in which every area of space is described by a vector indicating its strength and direction.
Fission	Breaking apart of a heavy object into two or more lighter objects.
Flavor	Describes the different types of leptons or quarks. For example, for neutrinos, the, flavors are electron, muon, and tau.
Fungi	Life forms characterized by an absence of chlorophyll, for example, mushrooms, yeasts, and molds.
Fusion	Combining two lighter constituents to make a heavier object.
Galaxy	Collection of stars that constitute the neighbors of the Sun and move in concordance with it. Our galaxy, the Milky Way Galaxy, contains 100–400 billion stars and is 100,000 light-years in size. The distance to the nearest adjacent galaxy, the Andromeda Galaxy, is 2.5 million light-years.
Gamma-rays	Electromagnetic radiation, or photons, that are at the high-energy end of the electromagnetic energy spectrum.
Gauss	Measure of the strength of a magnetic field. Earth's magnetic field has a strength of roughly one Gauss.
Genes	Unit of heredity in a living organism.
Global chirality	Chirality that is consistently the same everywhere, that is, it does not vary from one place to another.
Green sulfur bacteria	Living organisms observed in the Spring in Yellowstone National Park.
Green non-sulfur bacteria	Living organisms observed in the Octopus Spring in Yellowstone National Park.
Ground state	Lowest energy state in which an entity—an atom or a nucleus—can exist.
Gravitational potential energy	Energy associated with position. For example, when you hold an object in the air, it has gravitational potential energy, which gets converted to kinetic energy, that is, the energy of motion, when you release the object.

Guanine	One of the nucleotide bases constituting DNA.
Guide star (real or artificial)	Star (if real) that is in the field of view of the telescope that is used to adjust the mirror so as to optimize the resolution. If the guide star is artificial, it is generated by a laser beam that excites atoms in Earth's atmosphere, which then appear as a star and which allow resolution optimization.
Hartree–Fock	Iterative approximation technique used in a variety of quantum mechanical calculations. In this approximation, a mean-field potential from other particles are used.
Hayabusa	Japanese space mission that landed on comet Itokawa and returned to Earth with samples from that comet.
Helicity	Direction of spin of a particle with respect to its momentum. Positive helicity means that the spin points more in the direction of momentum, while negative helicity means that the spin points more in the direction opposite the momentum.
Helium burning	Second stage of stellar evolution, in which, primarily, three helium nuclei are converted into a carbon nucleus and an additional helium nucleus can be added to make an oxygen nucleus.
High-Z Supernova Project	One of the projects to determine the parameters of the universe by using Type Ia supernovae as standard candles.
Homochiral	Having only one chirality. A homochiral medium is totally enantiomeric.
Hubble's Law	$v = HR$; it relates the velocity of recession of distant objects to their distance.
Hubble constant	Proportionality constant, H, in Hubble's Law, currently at about 70 km s^{-1} Mpc^{-1}.
Hubble Space Telescope	Telescope flown into space so as to circumvent the absorption of ultraviolet and infrared light by Earth's atmosphere, and also to avoid the resolution spoiling from the motion of Earth's atmosphere.
Hydrodynamics	Description of matter, through physics equations, as if it were a continuous fluid.
Hydrogen burning	First stage of stellar evolution, in which hydrogen-induced nuclear reactions convert four protons into a helium nucleus.
Hydrothermal vent	Vent in the ocean floor through which a variety of chemicals can be emitted.
Hyperbolic comet	Comet that is made of stuff not from our solar system. It would make only one pass through the solar system, on a hyperbolic orbit, which is not closed on itself, and so would originate outside the solar system, pass through it, and then return to extrasolar space. Thus, the material constituting the hyperbolic comet would not necessarily have originated from the same region of space as the material constituting the solar system.
Hyperthermophiles	Form of extremophile that can withstand very high temperatures, 80 °C to more than 120 °C.
Infrared light	Light that has a longer wavelength than visible (optical) light, that is, longer than 700 nm.

In phase	Describes two oscillations, for example, two components of the electric field, that reach their respective maxima and minima simultaneously.
Intracluster medium	Superheated gas near the center of a galaxy cluster. It contains mostly hydrogen and helium, and strongly emits X-rays.
Interstellar medium	Material that exists in the space between the stars.
Ions or ionized atoms	Atoms or molecules that are not electrically neutral by virtue of having lost one or more of the electrons that would make them electrically neutral.
Isotope	Isotopes of an element have nuclei with the same number of protons but differing numbers of neutrons. Examples could be some of the isotopes of tin: ^{112}Sn, ^{114}Sn, ^{115}Sn, ^{116}Sn, ^{118}Sn, ^{120}Sn; or the stable isotopes of oxygen: ^{16}O, ^{17}O, and ^{18}O.
Itokawa	Comet from which the Japanese space mission *Hayabusa* brought samples to Earth
Juan de Fuca ridge	Location of a hydrothermal vent field in the northeast Pacific Ocean.
Junk DNA	Sections of human DNA that do not fulfill any obvious physiological or replicative need.
Kelvin temperature	Measurement of temperature in which zero Kelvin is absolute zero. Room temperature on this scale is about 300 K. Water freezes at 273.15 K, and it boils at 373.15 K.
Kepler Space Telescope	Telescope that is primarily dedicated to finding extrasolar potentially habitable planets.
Kingdoms	Classification of living things that includes archaea, bacteria, and four classes of eukaryotes.
Laser Interferometer Gravitational-wave Observatory, LIGO	Incredibly sensitive set of three worldwide detectors designed to observe the gravitational waves emitted from extraordinarily energetic cosmic events such as black hole mergers and neutron star mergers.
Lepton conservation	One of the conservation laws of physics that says that in any reaction or decay process, the number of leptons–antileptons must be conserved.
Leptons	Particles that do not interact via the strong interaction, and so are not made up of quarks. They include electrons, muons, taus, neutrinos, and their antiparticles.
Life signals	Detectable entities or events that would indicate the existence of living beings.
Ligand	Basic molecular unit that can be used as part of a more complex molecule.
Light curve	Temporal evolution of the light output from any star—for our purposes, a supernova.
Light-year	Distance light travels in one year. This is 9.47 trillion kilometers, or 5.9 trillion miles.
Linearly polarized light	Light in which the electric field vector appears to oscillate in one direction (see Figure 5.2).
Lipids	Organic compounds that are one of the components of living cells. They are greasy and not soluble in water. Their function is, among others, energy storage.

Local chirality	Situation that exists when chirality is established in one place, but another chirality may exist at a different location. This would be the situation that would exist from chirality resulting from circularly polarized light.
Longitudinal polarization	Situation that exists when the spin of a particle is parallel or antiparallel to its direction of motion. This is the case for photons and highly relativistic electrons.
Lookback time	Astronomers' jargon for time measured backward from the present.
Lorentz force	Force that moving charges experience in a magnetic field. The Lorentz force is perpendicular to both the external magnetic field and to the velocity of the charge.
LUCA	Last Universal Common Ancestor; thought to be the living organism from which all life originated.
Magnetic field	Physical entity that affects moving charged particles and also, along with the electric field, describes the motions of light particles, or photons.
Magnetic dipole moment	Property of elementary particles, nuclei, and atoms that interacts with an external magnetic field to produce an energy shift. Macroscopic quantities also have magnetic moments, for example, a current loop.
Magnetic monopole	Magnetic equivalent of an isolated charge. If magnetic monopoles existed, magnetic field lines would not have to connect on themselves, but could, as with electric charges, radiate outward to infinity.
Magnetic susceptibility	Measure of how much an object becomes magnetized in the presence of an external magnetic field. Objects with a nonzero magnetic susceptibility will have a nonzero magnetization in the presence of an external field.
Magnetic substates	Energy states of atoms or nuclei, or their combination, that result from the interaction of their magnetic moments with a magnetic field. The number of magnetic substates of an atom that has a total angular momentum J will be $2J + 1$, and the projections of the vector $\boldsymbol{J}$ along the magnetic field direction will range from $+\boldsymbol{J}$ to $-\boldsymbol{J}$ in steps of 1 (in units of Planck's constant).
Magnetization	Net vector sum of the dipole moments of a population of dipoles. For a population of dipoles that can only point in two directions, this is taken to be the net magnetic dipole moment of this population, which is taken as a sum of dipole moments pointing in one direction minus the sum of dipole moments pointing in the opposite direction.
Magnitude	Way in which astronomers describe the brightness of the objects they observe. An object that is one magnitude brighter than a second object is 2.512 times as bright as the second object. An object that is two magnitudes brighter than a second object is $(2.512)^2 = 6.310$ times as bright as the second object.

Mass energy	Refers to a way of indicating masses by indicating them instead as the energy they contain according to Einstein's famous equation $E = mc^2$, where m is the mass of the object and c is the speed of light.
Membrane (of a cell)	Container that holds the components of the cell.
Metabolism	Processes by which our cells sustain life. These allow cells to grow and reproduce.
Meteorite	Meteoroid that penetrates Earth's atmosphere and hits the ground.
Meteoroid	Chunk of rock, and possibly ice, that exists as part of the interstellar medium.
MeV	Unit of energy. One MeV = one million electron volts; a useful unit in nuclear physics, since that is a typical nuclear energy. One erg = 0.625×10^6 MeV.
Microbial mat	Mat of living entities observed in the Octopus Spring in Yellowstone National Park.
Miller–Urey experiment	Famous experiment, conducted in the early 1950s, in which an electric discharge in the presence of a few simple chemicals was found to produce amino acids.
Mirror isomers	Molecules that are identical in every respect, for example, melting and boiling temperature, solubility, but which have opposite chirality.
Molecular clouds	Giant clouds of gas, dust, and stars that occur in the Galaxy.
Møller–Plesset perturbation theory	Theoretical approach to calculating electron distributions in molecules in which single-particle Hamiltonians are perturbed to include correlation effects.
Monte Carlo	Computational algorithm used to simulate phenomena in which many of the variables are stochastic in nature. The algorithm works by choosing random assortments of individual variables according to a reasonable (or real) distribution of those variables and calculating the results of each event in the simulation.
Mothra hydrothermal vent field	Hydrothermal vent field in the northeast Pacific Ocean.
Muon	Elementary particle. Muons can be either positively or negatively charged. They are about 260 times as massive as electrons or positrons. They have a spin of $\hbar/2$.
Murchison meteorite	Meteorite that landed in 1969 in Australia that contained a variety of amino acids, all of which either had a left-handed chirality or were achiral.
Mutation	Change in the DNA coding in a cell that makes it different from what it previously was through, for example, inexact replication or cosmic rays.
Nebula	Cloud of gas and dust that is produced when a massive star explodes.
Neon burning	Stage of stellar evolution that is between carbon burning and oxygen burning.

Neutral-current weak interactions	Weak interactions that can transfer energy to a nucleus, but do not of themselves change the proton and neutron number of the nucleus.
Neutrino	Elementary particle that has no charge, a very tiny mass, and interacts only through the weak interaction. Neutrinos come in three flavors: electron, mu, and tau. Antiparticles exist for each of those flavors. They have a spin of $\hbar/2$.
Neutrino oscillations	Conversions between neutrinos of different flavors, due to the fact that the different flavors of neutrinos can change back and forth, or in some cases, just convert from one to another.
Neutron	One of the basic constituents of atomic nuclei and of neutron stars. Neutrons are composed of three quarks, one up quark and two down quarks. They have a spin of $\hbar/2$.
Neutron closed shells	Filling of the quantum mechanical shells in a nucleus; these exist for nuclei that contain 2, 8, 20, 28, 50, 82, or 126 neutrons.
Neutron star	Star that has collapsed essentially to the density of nuclei and is made up mostly of neutrons and a small fraction of protons. A neutron star is supported by the degeneracy pressure of its neutrons and protons.
Nuclear excited state	State of existence of a nucleus at a higher energy, that is, it is less tightly bound, than the ground state, or the lowest energy state.
Nuclear magnetic resonance (NMR)	Process in which material placed in a strong external magnetic field has its magnetic moment precess about this field. A resonant condition is set up when another oscillating magnetic field causes a significant portion of the molecules to align in the same direction if the oscillating field oscillates at the precession frequency or a multiple of the precession frequency.
Nuclear reactions	Interactions between two nuclei that result in different nuclei, either as a composite of the first two or as two or more new nuclei.
Nucleic acids	See DNA and RNA.
Nucleobases	Building blocks of RNA and DNA: adenine, cytosine, guanine (both DNA and RNA), thiamine (DNA only), and uracil (RNA only).
Nucleosynthesis	Creation of elements via nuclear reactions. This mostly takes place in stars.
Octopus Spring	A geothermal spring in Yellowstone National Park, located a few miles from the Old Faithful geyser.
Olber's paradox	Assertion that the night sky ought to be bright if the universe is infinite and uniformly populated with stars.
Oligomers	Molecules that consist of several smaller molecules joined together. By contrast, polymers can have an unlimited number of smaller molecules.
Optical light	Light having wavelengths that can be seen by the unaided eye, that is, with wavelengths from 400 to 700 nm.
Organic molecules	Generally, molecules that contain carbon. However, there are a few exceptions of molecules that contain carbon that are classed as inorganic molecules.

Oxygen burning	Stage of stellar evolution that occurs between neon burning and silicon burning.
Panspermia hypothesis	Hypothesis that life was created in outer space, or at least somewhere else, and then transported to Earth.
Paramedium bursaria	Very unusual pond-dwelling paramecium.
Paramagnetic	Having a (generally small) positive, nonzero magnetic susceptibility. Objects that are paramagnetic will have a net magnetization that points in the direction of an external magnetic field. These objects are attracted to other magnetic objects.
Parity	Handedness. Parity is also a fundamental symmetry of physics that is violated by the weak interaction, but is conserved in most cases.
Parity transformation	Transformation in which the coordinate system is reversed. This can also be accomplished by reversing the direction of one set of axes followed by a rotation of 180°.
Parsec	Astronomical distance unit = 3.26 light-years.
Pauli principle	"Law" of physics that requires that some types of particles in some systems, for example, electrons in atoms, or protons and neutrons in nuclei, have only one such particle in each allowed quantum mechanical state. This produces the well-defined sizes of atoms, nuclei, white dwarfs, and neutron stars.
Peptides	Relatively short protein-like chains of amino acids, but with different terminations from proteins.
Period of oscillation	Length of time for an oscillation to complete one full cycle.
pH	Indicator of acidity or basicity of a medium. Zero is essentially purely acidic, seven is neutral, and 14 is essentially purely basic. From the definition of pH, however, it is possible to have negative pH and pH greater than 14.
Photon	A fundamental "particle." This is the particle of electromagnetic energy and includes radio waves (very low energy), light (moderate energy), X-rays (higher energy), and gamma-rays (extremely high energy).
Photosphere	Periphery of a star.
Photosynthesis	Production of complex organic materials, especially carbohydrates, from carbon dioxide, water, and minerals using sunlight as an energy source.
Pink filamentous	Characteristics of a community of life found in Octopus Spring.
Pink streamers	Description of a community of life found in Octopus Spring.
Pion	Strongly interacting particle that is comprised of a quark–antiquark pair. There are three charge states: positive, neutral, and negative. They have a spin of zero.
Planck's constant	Unit in which microscopic angular momenta are measured. Its value is 6.626×10^{-27} erg s. It is also the unit that occurs in the famous Heisenberg Uncertainty Principle, which states that there is a fundamental limitation in the accuracy to which certain pairs of variables, for example, position and momentum, can be simultaneously measured.
Planck cosmological mission	A successor to *WMAP*; it measured the fluctuations in the 2.7 K cosmic microwave background radiation.

Planetary nebula	Glowing shell of ionized gas and plasma ejected during the asymptotic giant branch (helium-burning) phase of certain types of stars.
Polar vector	Vector that changes sign under a parity transformation.
Polarizability	Degree to which a material can be polarized. In the case of nuclear spins, it is the degree to which all spins can be aligned.
Polarized light	Light that has a specific orientation for its electric field vector (linearly polarized light) or an electric field vector that rotates in direction over time (circularly or elliptically polarized light).
Polycyclic aromatic hydrocarbons	Complex molecules (generally more complex than amino acids) observed in outer space.
Polymers	Long molecules that consist of shorter segments called monomers.
Positron	Elementary particle that is the antiparticle of the electron. It has the same mass as the electron, but has a positive charge, and has a spin of $\hbar/2$.
pp-chain	Set of reactions that defines how hydrogen nuclei get fused into helium nuclei in not-too-massive stars.
Primary process	Process of nucleosynthesis that creates a new set of seed nuclei each time it operates, and therefore produces the same results in stars no matter how many heavy nuclei they had previously. The r-process is primary; the s-process is not.
Progenitor star	Star that existed prior to a supernova explosion.
Prokaryotes	Not eukaryotes, that is, archaea or bacteria.
Protective mechanisms	Means by which extremophiles shield themselves from the extreme environments in which they live.
Protein	Complex organic molecules on which we depend to sustain our lives. They are made from amino acids.
Protein world	View that the first molecules from which life eventually evolved must have been proteins, since making molecules as complex as RNA would have been too unlikely.
Protista	Unicellular protozoans and multicellular algae.
Proton	One of the basic constituents of atomic nuclei, and to a small extent, of neutron stars. Protons are comprised of three quarks: two up quarks and one down quark. They have a spin of $\hbar/2$.
Pseudoscalar	Scalar value that changes sign under a parity transformation, such as helicity.
Quantum mechanics	Theory that describes microscopic nature. It results in the quantization of many entities that exist in steady-state situations, such as energy and angular momentum, in atoms and nuclei.
Quantum state	Characterization of all the properties of existence for a particle, including spatial distribution, momentum, energy, angular momentum, spin, and whatever other variables are needed to completely specify the state.
Quark	A fundamental particle. Quarks cannot be found by themselves, but either occur as three quarks together in particles like protons and neutrons, or as a quark–antiquark pair in mesons like the pion. Quarks have a spin of $\hbar/2$.
Racemic	Referring to an assembly of chiral molecules that has equal populations of the two possible chiral states.

Radiation pressure	Pressure created by the outflow of photons from the center of a star.
Radicals	Atoms, molecules, or ions with unpaired electrons or an open electronic shell. They may be positively or negatively charged, or neutral.
Radiocarbon dating	Use of the radioactive nucleus ^{14}C (half-life = 5730 years) to determine the ages of objects, e.g., wood and bone, that contain carbon. This technique only works for ages up to a maximum of 100,000 years.
Radio waves	Electromagnetic radiation that has long wavelengths (greater than 1 mm) and correspondingly low energies.
Rankine	One of several temperature scales. On this scale, absolute zero is 0 °R, and water freezes at 491.67 °R and boils at 671.67 °R.
Reaction probability	Likelihood that a reaction will occur. See cross section.
Red giant	Star that arises during helium burning, when the core of the star generates more energy than it previously had, and the radiation pressure forces the periphery of the star outward. This results in a larger surface area that, since the energy emitted by each unit area is reduced, appears redder than the star was in its hydrogen-burning phase.
Redshift	Shift that occurs in spectral lines from atoms or ions resulting from the source of those lines moving away from the observer.
Repair mechanisms	Means by which cells repair components, especially DNA, that have been damaged by the extreme environments in which they live.
Replication	Reproduction; this applies to living beings, cells, DNA, etc.
Ribosome	Site within a cell where protein assembly from the amino acid constituents occurs.
RNA	Ribonucleic acid. One of a set of molecules that are involved in carrying on the functions of life, including assembling the proteins in cells. They include messenger RNA, or mRNA, and transfer RNA, or tRNA.
RNA world	View that the first molecules from which life evolved must have been RNA, since without RNA the proteins do not have their instructions for formation.
Rosetta	The European Space Agency mission that sent a lander to the comet Churyumov Gerasimenko to analyze samples.
r-process	Process of nucleosynthesis that involves an extremely rapid sequence of neutron captures, synthesizing half of the nuclides heavier than iron, and all nuclides heavier than ^{209}Bi.
SETI	Search for Extraterrestrial Intelligence. This usually refers to the search for radio signals that would characterize extraterrestrial intelligent life.
Shielding	Shift in the magnetic field at a nucleus in a molecule due to the motion of the orbital electrons in the molecule. Shielding tends to reduce the overall local magnetic field at the nucleus.
Shock wave	Propagating disturbance in a solid, gas, or plasma. When a star collapses to slightly more than nuclear density, it first overshoots

	that density, then bounces back, thereby producing an outward-going shock wave in the material that is external to it.
Silent supernovae	Core-collapse supernovae that ultimately collapse to a black hole, although they might first collapse to a neutron star. However, in either case, the black hole will swallow most of the electromagnetic radiation before it can escape the star. Neutrinos, however, will be emitted, and will escape before the black hole is formed.
Silicates	Inorganic molecules that contain silicon.
Silicon burning	Final phase of stellar evolution, which follows oxygen burning and precedes the collapse to a neutron star or black hole.
Smectite clay	Clays are built of two types of sheets, each defined by its chemical structure. Smectite clays have a sheet of one type sandwiched between two sheets of the other type.
SNAAP model	Supernova Neutrino Amino Acid Processing model. It describes how amino acids achieved their chirality.
Solar system	System of planets and our Sun.
Spallation neutron source	Facility for producing intense neutron beams, located in Oak Ridge, Tennessee. It may also be used to produce intense neutrino beams.
Spectrograph	Instrument that allows astronomers to resolve the different wavelengths of light emitted by a star. These can be used to identify the abundances of the elements in the photosphere of the star.
Spectrum	Distribution of energies of the electromagnetic radiation over its possible range. This can also be the distribution over the range of wavelengths. Spectrum can also refer to a distribution of energies of particles. The plural of spectrum is spectra.
Spin	An intrinsic degree of freedom of particles and nuclei that is a form of angular momentum.
Spitzer Space Telescope	Spaceborne observatory of infrared radiation. It is sensitive to light of wavelengths that are absorbed by Earth's atmosphere, hence the reason it is in space. It was named after Lyman Spitzer, Jr.
Spores	Single-walled or multiple-celled reproductive bodies of organisms that are capable of surviving for long periods in hostile conditions.
s-process	Process of nucleosynthesis that involves a slow (compared to typical β-decay half-lives) sequence of neutron captures and β-decays, synthesizing half of the nuclides heavier then iron and terminating at ^{209}Bi.
Standard candle	Star that has some feature that allows astronomers to make an absolute determination of its distance. Examples are Cepheid variables, which have an oscillating luminosity that is related to its absolute luminosity, and Type Ia supernovae, all of which have approximately the same luminosity.
Standard solar model	Model that describes the details of operation of the Sun. This includes all of the complex hydrodynamics necessary in this

	description, as well as the nuclear reactions, photon and neutrino opacities, and other features necessary for a complete description.
Starburst superwinds	Winds that are generated from regions with an unusually high density of supernovae—the starburst regions. Superwinds are several times greater in mass, however, than could be generated by supernovae alone.
Stardust	NASA mission to sample comet 81P/Wild 2 and return to Earth with some of its material.
Steady-state universe	Cosmological theory that the universe had no beginning, but operated in a steady state. This theory is no longer viable.
Stellar evolution	Different stages of nuclear burning that exists in massive stars, along with the concurrent hydrodynamic conditions that exist.
Stellar winds	Winds by which some stars expel their outer layers into the interstellar medium. These are often the result of radiation pressure from the photons being produced in the star.
Strong force/interaction	One of the basic forces or interactions of physics.
Supernovae	Stellar explosions that produce so much light that they can outshine entire galaxies. Type Ia supernovae are powered by thermonuclear processes and blow up the entire star. Type II supernovae are core-collapse supernovae and are powered by gravity. They produce a neutron star or a black hole. Type Ib and Ic supernovae are also core-collapse supernovae, but they have shed their outer hydrogen, and perhaps helium, envelopes before they explode.
Supernova Cosmology Project	Program that measured Type Ia supernovae at a range of distances. The data from the SCP demonstrated the existence of dark energy.
Supernova HO for the Equation of State (SHOES)	Project to determine the parameters of the universe using several types of standard candles, including Type Ia supernovae. The *Hubble Space Telescope* was used to observe them.
Supernova remnants	Made up of the gas that contains the newly synthesized elements that the supernova expels into the interstellar medium.
Tau	Fundamental particle that can be either positively or negatively charged and is more massive than muons. Taus have a spin of $\hbar/2$.
Template	In this context, a molecular segment that provides a pattern for more molecules to follow.
Tensor	Mathematical construct with "dimensionality." A tensor is specified by its rank. A rank-0 tensor is a scalar—a single number. A rank-1 tensor is a vector. A rank-2 tensor is a two-dimensional object, or a matrix.
Terrestrial	Of Earth.
Thermal energy	Energy associated with the motion of particles in a medium. This is heat energy.
Thermonuclear runaway	Condition wherein the heat generated by nuclear processes in a star cannot be compensated by the expansion of the star, which would cool it. This is what causes a Type Ia supernova to explode.
Thermophiles	Type of extremophile that lives at fairly high temperatures: 60 °C–80 °C.

Thermophilic cyanobacterium	One of the life forms observed in Octopus Spring in Yellowstone National Park.
Threshold energy	Energy that must be supplied to a reaction in order for it to proceed.
Thymine	One of the nucleotide bases constituting DNA.
Time dilation	Result of moving at speeds close to the speed of light, wherein the decay lifetimes of particles can become much longer, and lengths contract.
Time projection chamber (TPC)	Device used to identify particles and their energy by examining their tracks in a gas-filled detector. Tracks are mapped by measuring the time it takes their ionization trails to travel to a detector plane.
Time reversal transformation	Transformation in which the sign of the time quantities is reversed. That is, $t \rightarrow -t$, $\Delta t \rightarrow -\Delta t$, and $\partial t \rightarrow -\partial t$.
Type Ia supernova	Stellar explosion in which a white dwarf exceeds its maximum mass due to accretion of matter from a companion. This results in thermonuclear runaway.
Ultraviolet light	Light having wavelengths shorter than those in the optical; light with a wavelength less than 400 nm (but not as energetic as X-rays).
Universe	Collection of all the stars and galaxies that we believe exist.
Uracil	One of the nucleotide bases.
Vector	Pictorial representation of a physical quantity. It has both length and direction, with the length representing the strength of the quantity being represented.
Visible light	Light in the wavelength range from about 400 nm to 700 nm and visible to the naked eye.
Wavefunction	Quantum mechanical description of a particle describing the probability that the particle will be at some point in space with some momentum. Wavefunctions are also used to describe the probability that a particle has a specific set of quantum numbers.
Wavelength of an oscillation	Distance over which an oscillation completes one cycle.
Weak interaction	One of the basic interactions of physics. The weak interaction mediates the process of nuclear β-decay, as well as the interactions of neutrinos with nuclei and other particles.
White dwarf	Final state of a medium-mass star in which the size is maintained by electron degeneracy pressure. These are made up either of carbon and oxygen, or of oxygen and magnesium.
White smokers	Hydrothermal vents that emit liquid carbon dioxide.
Wild 2	Technically 81P/Wild 2; it is the comet visited by the NASA mission *Stardust* and found to contain amino acids.
Wilkinson Microwave Anisotropy Project	*WMAP* measured the fluctuations in the 2.7 K cosmic microwave background radiation and obtained the most accurate (at the time its data analysis was completed) data on the Hubble constant, the baryonic and dark matter densities of the universe, and the component of dark energy.

Wind-blown bubbles	Bubbles that result from the winds from massive (greater than 8 solar mass) stars.
Wolf–Rayet stars	Very hot massive stars that have shed either one or two of their outer layers. If the former, they will be characterized by nitrogen emission lines. If the latter, they will exhibit carbon and oxygen emission lines.
X-rays	Electromagnetic radiation that is more energetic than ultraviolet light but less energetic than gamma-rays.
Yamanote line	Mass-transit train line that encircles Tokyo.
Zircons	Zirconium silicate, $ZrSiO_4$. This is produced in magma. Zircons are crucial for determining the ages of rocks because they initially contain uranium, but not lead, to which the uranium decays. This makes it possible to determine the ages of zircons.
Zwitterion	Neutral molecule with separate positively and negatively charged groups. Zwitterions can have a charge distribution created by atoms moving from one point of the molecule to another.

CPSIA information can be obtained
at www.ICGtesting.com
Printed in the USA
LVHW071825140319
610514LV00020BA/344/P